Advanced Technologies for Sustainable and Effective Waste Management Systems

Advanced Technologies for Sustainable and Effective Waste Management Systems

Edited by

Harshita Jain and Maulin P. Shah

IWA PUBLISHING

Published by **IWA Publishing**
Unit 104–105, Export Building
1 Clove Crescent
London E14 2BA, UK
Telephone: +44 (0)20 8054 8202
Email: publications@iwap.co.uk
Web: www.iwaponline.com

First published 2026

British Library Cataloguing in Publication Data
A CIP catalogue record for this book is available from the British Library

ISBN: 9781789065398 (paperback)
ISBN: 9781789065404 (eBook)
ISBN: 9781789065411 (ePub)

Doi: 10.2166/9781789065404

This eBook was made Open Access in September 2026.

Contents

Dedication

This book is dedicated to the relentless pursuit of knowledge and innovation in the field of environmental sustainability.

We dedicate this work to the researchers, educators, engineers, and policymakers around the world who are committed to developing sustainable solutions for waste management and environmental protection.

We also extend our heartfelt gratitude to our mentors, colleagues, students, and families whose continuous encouragement, inspiration, and support have made this work possible.

May this book contribute, in its own small way, to advancing sustainable technologies and responsible environmental stewardship for present and future generations.

About the Editors

Harshita Jain
PhD (Environmental Sciences)
Assistant Professor, Department of Climate Change and Environmental Sustainability, Amity Institute of Environment and Sustainability, Amity University Uttar Pradesh (AUUP), Gautam Budh Nagar, Noida, Uttar Pradesh 201303, India
Email: hjain@amity.edu

Dr Harshita Jain is an early-career young academic researcher with a background in membrane separation technology, sustainable development, pollution management, and water and wastewater treatment. She has been involved in numerous academic and research activities, including the publication of articles in SCI/SCIE journals. Innovations in clean technology, water and wastewater treatment, biosorption, vermicomposting, antibiotic resistance, public health, machine learning in disaster response, groundwater remediation, the impact of climate change, and artificial intelligence solutions for disaster preparedness are among her areas of interest. Dr Jain has also served as an editor for reputable books and regional textbooks in the sciences. Additionally, she manages multiple corporate social responsibility projects for major companies and has successfully completed various government-funded training programs. She has demonstrated a commitment to faculty development and community involvement by participating in initiatives that support innovative teaching strategies and sustainable development. Furthermore, she contributes significantly to the academic community by supervising PhD students and offering guidance for ME/MTech dissertations.

Maulin P. Shah
Adjunct Faculty, Lovely Professional University, Phagwara, Punjab, India
Email: maulinshah1979@gmail.com

Maulin P. Shah has been an active researcher and scientific writer in his field for over 20 years. He received a BSc (1999) in microbiology from Gujarat University, Godhra (Gujarat), India, and earned his PhD (2005) in environmental microbiology from Sardar Patel University, Vallabh Vidyanagar (Gujarat) India. His research interests include biological wastewater treatment, environmental microbiology, biodegradation, bioremediation, and phytoremediation of environmental pollutants from industrial wastewaters. He has published more than 250 research papers in national and international journals of repute on various aspects of microbial biodegradation and bioremediation of environmental pollutants. He is the editor of 200 books of international repute.

List of Contributors

Asif, Sk Md
Department of Agriculture, Brainware University, Barasat, Kolkata 700125, India. Email: skmdasifbckv@gmail.com

Bhati, Niketa
Amity Institute of Environmental Sciences, Amity University, Noida, Uttar Pradesh 201301, India

Cetinkaya, Turgay
Department of Aquatic Biotechnology and Genomics, Faculty of Aquatic Sciences, Istanbul University, Istanbul 34134, Türkiye. Email: turgaycetinkaya@istanbul.edu.tr

Dadhich, Sushmita M.
Division of Soil and Water Conservation Engineering, FOAE, SKUAST-Jammu, Jammu, India. Email: sushmitam52@skuastj.org

Diksha
Division of Environmental Science, ICAR–Indian Agricultural Research Institute, Pusa Campus, New Delhi 110012, India

Dutta, Ditimoni
Department of Food Technology, Parul Institute of Applied Sciences, Parul University, Vadodara 391760, India. Email: ditimoni.dutta42805@paruluniversity.ac.in

Fatima, Shahin
Department of Computer Science and Engineering, Koneru Lakshmaiah Education Foundation, Hyderabad 500075, Telangana, India. Email: shahin.fatima@klh.edu.in

Hazra, Sumit
Department of Computer Science and Engineering, Koneru Lakshmaiah Education Foundation, Hyderabad 500075, Telangana, India. Email: sumit.hazra@klh.edu.in

Hossain, Akbar
Soil Science Division, Bangladesh Wheat and Maize Research Institute, Dinajpur 5200, Bangladesh

Jagadeeswaran, R.
Department of RS & GIS, Tamil Nadu Agricultural University, Coimbatore, India

Jain, Harshita
Department of Climate Change and Environmental Sustainability, Amity Institute of Environment and Sustainability, Amity University Uttar Pradesh (AUUP), Gautam Budh Nagar, Noida, Uttar Pradesh 201301, India. Email: hjain@amity.edu

Jaswal, Shivani
Department of Food Science and Technology, Dr. Y. S. Parmar University of Horticulture and Forestry, Solan 173230, India. Email: shivanijaswal0205@gmail.com

Jindal, Suhasini
Amity Institute of Environmental Sciences, Amity University, Noida, Uttar Pradesh 201301, India

Joshi, Varad
Department of Computer Science and Engineering, Koneru Lakshmaiah Education Foundation, Hyderabad 500075, Telangana, India. Email: varadj40@gmail.com

Karimidastjerd, Atefeh
Food Engineering Department, Faculty of Chemical and Metallurgical Engineering, Yildiz Technical University, Istanbul, Türkiye. Email: atefeh.dastjerd@yildiz.edu.tr

Kaur, Sukhmandeep
Department of Food Science and Technology, Punjab Agricultural University, Ludhiana 141001, Punjab, India. Email: sukmandeepkaur1999@gmail.com

Khan, Aftab Ahmad
Department of Civil and Environmental Engineering, College of Engineering, King Faisal University, Al-Ahsa 31982, Saudi Arabia. Email: aabkhan@kfu.edu.sa

Kumari, Maya
Amity School of Natural Resources and Sustainable Development, Amity University Noida, Uttar Pradesh, India

Mandal, Argha
Department of Soil Science and Agricultural Chemistry, Visva-Bharati University, Bolpur 731235, India

Mohan V., Aparna
ICAR–Central Tuber Crop Research Institute, Sreekariyam, Thiruvananthapuram, Kerala, India

Munzeiwa, Wisdom Archford
Chemistry Department, Bindura University of Science Education, Bindura, Zimbabwe, Email: wmunzeiwa@buse.ac.zw

Oladipo, Iyabo Christianah
Department of Science Laboratory Technology (Microbiology Unit), Ladoke Akintola University of Technology, Ogbomoso, Nigeria. Email: icoladipo@lautech.edu.ng

Pal, Indrajit
Asian Institute of Technology, Disaster Preparedness, Mitigation and Management Program, Thailand

Raju, M.
Department of Agronomy, Tamil Nadu Agricultural University, Coimbatore, India

Sabharwal, Gourav
Department of Agronomy, Tamil Nadu Agricultural University, Coimbatore, India. Email: sabharwal308@gmail.com

Selvakumar, S.
Center for Water and Geospatial Studies, Tamil Nadu Agricultural University, Coimbatore, India

Sen, Arup
Department of Agricultural Chemistry and Soil Science, Bidhan Chandra Krishi Viswavidyalaya, Nadia 741252, India

Shah, Maulin P.
Adjunct Faculty, Lovely Professional University, Phagwara, Punjab, India

Sharma, Vivek
Agricultural and Biological Engineering Department, University of Florida, Gainesville, FL 32611, USA

Shrivastava, Anamika
Amity Institute of Environmental Sciences, Amity University Uttar Pradesh, Noida, India

Siddique, Shadab
Department of Computer Science and Engineering, Koneru Lakshmaiah Education Foundation, Hyderabad 500075, Telangana, India. Email: cseshadabsiddiqui@gmail.com

Singh, Lovepreet
Department of Chemical Engineering, Thapar Institute of Engineering and Technology, Patiala 147004, India. Email: lovepreet.singh@thapar.edu

Ucbaş, Derya
Department of Food Engineering, Eskişehir Osmangazi University, Turkey. Email: derya.ucbas@ogu.edu.tr

Victor, Agbogo Ugbetan
Department of Chemistry, Nigerian Army University Biu, Nigeria. Email: victor.agbogo@naub.edu.ng

Vaiyapuri, K.
Department of Agronomy, Tamil Nadu Agricultural University, Coimbatore, India

Foreword

Sustainable waste management has become one of the most pressing environmental priorities of the twenty-first century. Rapid urbanization, industrial development, and population growth have significantly increased the generation of wastewater and solid waste across the globe. Traditional waste management practices, which largely focused on disposal, are no longer adequate to address the environmental, economic, and societal challenges associated with modern waste streams. Instead, the focus is gradually shifting toward innovative, technology-driven, and resource-efficient approaches that align with the principles of sustainability and the circular economy.

Advanced Technologies for Sustainable and Effective Waste Management Systems provides a timely and comprehensive exploration of these emerging approaches. The book brings together contributions from an international group of researchers and experts who present cutting-edge developments in wastewater management, advanced treatment technologies, digitalization, policy frameworks, and sustainable implementation strategies. The diversity of perspectives reflected in the chapters highlights the interdisciplinary nature of modern environmental management and underscores the need for collaborative solutions.

One of the key strengths of this book is its balanced integration of technological innovations with governance and policy considerations. While several chapters explore advanced treatment processes, smart wastewater collection systems, and emerging digital tools such as artificial intelligence and data analytics, others address nature-based solutions, energy recovery strategies, and spatial technologies like remote sensing and GIS for waste management planning. Together, these topics reflect the evolving landscape of environmental technologies that are transforming waste from an environmental burden into a valuable resource.

The book also emphasizes the importance of integrating technological progress with effective policy frameworks and practical implementation strategies. Case studies and global perspectives included in the volume provide valuable insights into real-world applications and the challenges faced in translating innovative concepts into operational systems. Such discussions are particularly relevant for developing regions where sustainable infrastructure and efficient waste management systems are urgently needed.

This book will serve as a valuable reference for researchers, environmental scientists, engineers, policymakers, and students who are working toward sustainable solutions for waste and wastewater management. By presenting both conceptual advancements and practical approaches, the volume contributes to bridging the gap between research, technology development, and real-world implementation.

I commend the editors and contributing authors for their efforts in bringing together this comprehensive and insightful work. It is my hope that this book will stimulate further research, foster interdisciplinary collaboration, and support the development of sustainable waste management systems that protect environmental resources and promote long-term societal well-being.

Professor P. R. Shah (Retired)
Dept of Microbiology, Gujarat University, Gujarat, India

doi: 10.2166/9781789065404_xxiii

Preface

Rapid urbanization, industrial expansion, and population growth have significantly increased the generation of municipal and industrial wastes worldwide. Among these, wastewater and solid waste present some of the most pressing environmental challenges, particularly in rapidly developing regions. Conventional waste management systems, which were primarily designed for disposal rather than resource recovery, are increasingly inadequate to address the complexities of modern environmental sustainability. Consequently, there is an urgent need to transition toward integrated, technology-driven, and circular approaches to waste management that emphasize efficiency, environmental protection, and resource valorization.

Advanced Technologies for Sustainable and Effective Waste Management Systems aims to provide a comprehensive overview of emerging technologies, innovative management strategies, and policy frameworks that can support sustainable waste management practices. This edited volume brings together contributions from researchers and practitioners across multiple disciplines and countries, offering insights into both theoretical developments and practical applications in the field.

The chapters collectively explore a wide spectrum of topics related to smart wastewater management, advanced treatment technologies, resource recovery, digitalization, and governance frameworks. The opening chapters focus on the evolution of wastewater management within the context of smart and circular systems, highlighting the importance of integrated infrastructure and modern monitoring approaches. Subsequent chapters examine technologies for smart wastewater collection and advanced treatment processes, including intelligent systems that enhance treatment efficiency and environmental performance.

Several chapters also address nature-based solutions and resource recovery strategies, such as phytoremediation using aquatic macrophytes and innovative energy recovery mechanisms from wastewater streams. These approaches reflect the growing emphasis on transforming waste into valuable resources, thereby supporting circular economy principles and sustainable agricultural applications.

Recognizing the increasing role of digital transformation in environmental management, the book also explores data analytics, digitalization, artificial intelligence, and decision-support systems for optimizing wastewater management operations. The inclusion of topics such as federated learning and edge artificial intelligence for smart city wastewater systems highlights the convergence of environmental science with advanced computational technologies.

Beyond technological innovations, sustainable waste management also requires effective governance and institutional support. Therefore, there are chapters dedicated to policy frameworks, governance strategies, and global case studies, providing insights into implementation challenges and future directions. Additionally, the integration of remote sensing and GIS-based approaches for solid waste disposal site suitability mapping demonstrates how spatial technologies can enhance planning and decision-making processes.

This book is intended for researchers, environmental engineers, policymakers, practitioners, and graduate students working in environmental science, water resources management, and sustainable infrastructure development. By bringing together interdisciplinary perspectives and international experiences, the volume seeks to contribute to the advancement of sustainable and resilient waste management systems.

The editors express their sincere gratitude to all contributing authors for their valuable efforts and scholarly contributions. We also acknowledge the support of reviewers, colleagues, and institutions that helped make this book possible. It is our hope that it will serve as a useful reference for advancing research, innovation, and policy development in sustainable waste management.

Acknowledgments

The successful completion of this book would not have been possible without the valuable contributions, encouragement, and support of many individuals and institutions. The editors would like to express their sincere gratitude to all the contributing authors who dedicated their time and expertise to prepare high-quality chapters for this volume. Their scholarly insights and interdisciplinary perspectives have significantly enriched the content and scope of this book.

We also extend our appreciation to the reviewers and academic peers who provided constructive suggestions and critical feedback during the development of the manuscript. Their thoughtful comments helped improve the clarity, depth, and overall quality of the chapters.

The editors gratefully acknowledge the support of their respective institutions for fostering an environment that encourages academic research and collaboration. In particular, we appreciate the academic support and encouragement provided by Amity Institute of Environmental Sciences, Amity University, Noida, which has consistently promoted research and innovation in environmental science and sustainable development.

We would also like to thank our colleagues, collaborators, and students whose discussions, ideas, and encouragement contributed in many ways to the development of this work. Their enthusiasm and engagement continue to inspire research efforts in the field of sustainable waste management.

Finally, we express our heartfelt gratitude to our families for their constant encouragement, patience, and understanding throughout the process of compiling and editing this book. Their support has been invaluable in bringing this work to completion.

It is our hope that this book will contribute to advancing knowledge, innovation, and practical solutions in sustainable waste management and environmental protection.

doi: 10.2166/9781789065404_0001

Chapter 1

Sustainable wastewater management in the era of smart and circular systems

Harshita Jain[1,*], *Maulin P. Shah*[2] *and Niketa Bhati*[1]

[1]*Department of Climate Change and Environmental Sustainability, Amity Institute of Environment and Sustainability, Amity University Uttar Pradesh (AUUP), Noida, Gautam Budh Nagar, Uttar Pradesh 201303, India*

[2]*Adjunct Faculty, Lovely Professional University, Phagwara, Punjab, India*

**Corresponding author: hjain@amity.edu*

ABSTRACT

In an age of increasing urbanization, industrialization, and climate change, water management has become a crucial issue to contemporary environmental governance. Traditional wastewater systems that are primarily aimed at dumping and regulatory compliance are not addressing the interconnected environmental, social, and resource challenges of the present. This chapter introduces wastewater management as a rapidly changing practice which today intersects sustainability, circular economy, and digital innovation. In this chapter we discuss how wastewater is increasingly viewed as an economic rather than an incurable burden and provides opportunities for energy, nutrients, and water recovery. It examines how systems that are progressively shifting away from linear treatment strategies toward intelligent, adaptive, and data-informed systems that improve operation efficiency and environmental resiliency. Monitoring technologies, system integration, and decision-support tools are considered for facilitating this shift. The chapter builds on the concept of wastewater management and, in so doing, the book's conceptualization by situating wastewater management within broad sustainability and circularity contexts. It also outlines the scope and organization of the subsequent chapters explaining how technological, ecological, and policy perspectives are layered to address current and potential problems of wastewater management.

Keywords: Sustainable wastewater systems, circular economy, smart infrastructure, resource recovery, urban water management, system resilience

1.1 INTRODUCTION

Wastewater management has been undergoing a dramatic change in the last decade primarily in response to the rapid urbanization and industrialization

(Dutta *et al.*, 2021). As cities become larger and more people move into the waterways, domestic sewage and industrial effluents flow more rapidly into treatment systems (Silva, 2023). Traditional practices of treatment of wastewater that were mainly aimed at collection and disposal have become insufficient to handle these growing demands (Öberg *et al.*, 2020). According to Rafique *et al.* (2022), industrial practices are also complex because they contain a wide range of chemical contaminants, heavy metals, and high strength organic waste that require advanced treatment methods. Plus, urbanization often increases faster than wastewater treatment, creating overloaded sewers, sewage discharges, and environmental degradation (Silva, 2023). The need to think about wastewater as not just waste, but as a resource that could be recovered and reused is consequent. Lehmann (2017) explains this perspective shift that highlights the need to consider integrated wastewater management approaches that combine urban development with environmental sustainability, public health protection, and resource efficiency (Figure 1.1).

Climate influences, and the decreasing supply of freshwater, further weaken the resilience of wastewater systems (Sweetapple *et al.*, 2018). Hyde-Smith *et al.* (2022) state that modifications to precipitation, extreme weather, and elevated temperatures can impede traditional wastewater treatment systems and result in overflows, reduced treatment efficiency, and downstream ecological impacts. Untreated, or partially treated, waste from poor regions is essentially reused for agricultural, industrial, or urban use and is therefore extremely reliable and safe. These challenges also highlight the need for resilient and adapted wastewater systems that can respond to environmental change (Li *et al.*, 2023a). It is an increasingly important concept to focus on

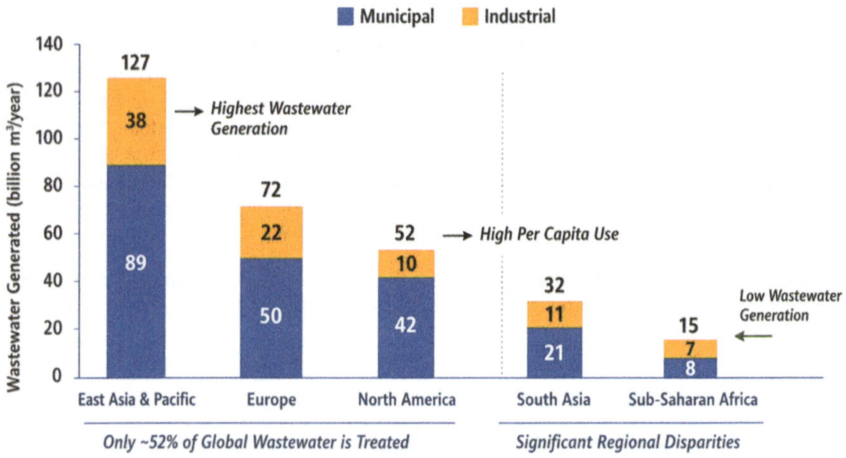

Figure 1.1 Global wastewater generation and urban growth (Source: Copernicus ESSD (2021), WHO SDG 6.3.1 Report (2024).)

smart monitoring, predictive modeling, and decentralized treatment plans to improve system robustness (Das & Mishra, 2025). Li *et al.* (2023b) state that by situating wastewater management within the larger theme of climate resilience and sustainable use of water, policymakers, engineers, and researchers can now develop approaches that will safeguard environmental and long-term availability of water resources.

1.2 GLOBAL TRENDS AND CONTEMPORARY CHALLENGES

Worldwide, population, urbanization, and intensified industrial activity are bringing with them the production of wastewater at a faster rate than ever before (Vs *et al.*, 2011). As per Ravina *et al.* (2021), recent analysis has revealed that the majority of cities in low- and middle-income countries cannot provide adequate sewage collection and treatment services, which generate significant gaps between wastewater generation and treatment capacity (Table 1.1). Even in some developed regions, aging infrastructure and capacity constraints result in periodic overflows and untreated discharges (Shah, 2024; Shah & Rodriguez-Couto, 2019). This misunderstanding between demand and infrastructure presents the urgent need for creative, scalable, and flexible treatment strategies that can handle growing volumes of wastewater while also protecting the environment (Figure 1.2).

Untreated or treated wastewater has serious environmental and human health consequences (Wu *et al.*, 2024). Fluvia from economically rich effluents create eutrophication in rivers and lakes disrupting aquatic life and decreasing biodiversity (Batuwita *et al.*, 2022). Industrial wastes can contain toxic chemicals, heavy metals, and persistent organic pollutants that pollute soil,

Table 1.1 Global trends in wastewater generation and treatment coverage.

Region/ Country	Population (millions)	Wastewater Generated (MLD)	Treatment Coverage (%)	Key Challenges
North America	370	85 000	95	Aging infrastructure, nutrient discharge
Europe	740	140 000	90	Seasonal variability, energy-intensive treatment
Asia	4600	450 000	60	Rapid urbanization, inadequate decentralized systems
Africa	1400	50 000	30	Limited infrastructure, funding constraints
Latin America	650	60 000	55	Informal settlements, industrial discharges

Note: Population and wastewater values are indicative; sources include WHO/UN reports.

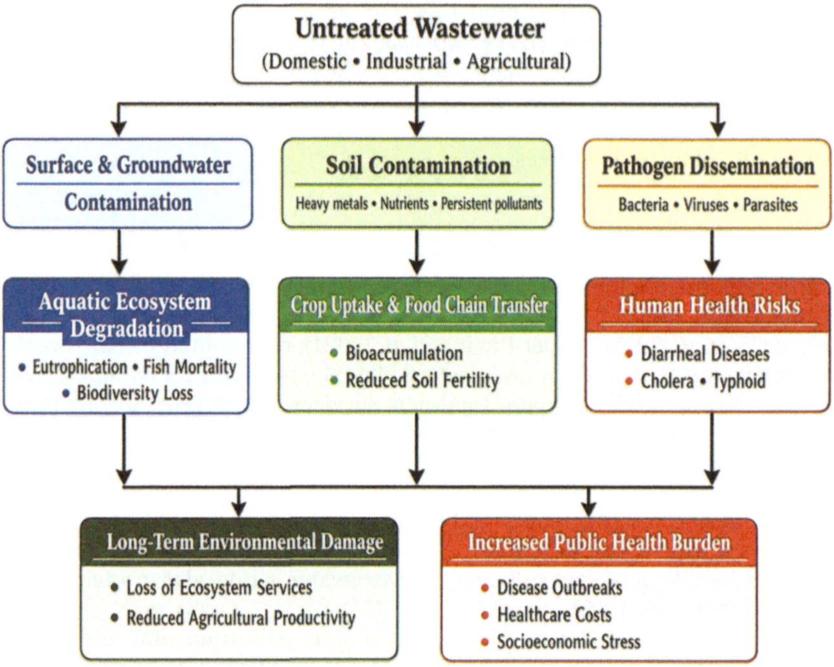

Figure 1.2 Environmental and public health implications of untreated wastewater.

groundwater, and surface water (Pastor & Hernández, 2011). On the human side, water, food, or recreational contact with pathogenic microorganisms and chemicals, increase the risk of waterborne diseases and chronic health conditions (Stec *et al.*, 2022). As per Guest *et al.* (2009), these environmental and public health pressures further emphasize the importance of more comprehensive approaches to wastewater treatment that transcend regulatory compliance and blend sustainability and resource recovery goals.

Traditional wastewater treatment systems, that would otherwise be based on activated sludge or lagoon treatment, face multiple challenges in meeting the challenges of today (Rezai & Allahkarami, 2021). These are often energy intensive, design rigid, and unable to cope with fluctuations in loads or new contaminants. According to Guest *et al.* (2009), resource recovery and reusing of energy, nutrients, and water is often overlooked in traditional models, and so the chances to participate in a circular economy are numbered. Plus, the flexibility to adapt to climate change-induced stresses or rapidly growing urban populations is limited (Hobbie & Grimm, 2020). As a result, wastewater management is shifting toward smart, adaptive, and integrated treatment systems that combine technological innovation, environmental stewardship, and operational efficiency (Garg *et al.*, 2024).

1.3 SUSTAINABILITY PRINCIPLES IN WASTEWATER MANAGEMENT

1.3.1 Environmental, economic, and social dimensions of sustainability

Wastewater sustainability goes beyond just treatment of effluent; it also includes environmental, economic, and social impacts (Muga & Mihelcic, 2007). As per Makanda *et al.* (2022), sustainable practices, either environmentally or physically, aim to reduce pollution, conserve water resources and promote healthy ecosystems. This includes minimizing chemical and energy consumption, reusing valuable nutrients, and safe reuse of treated water (Guest *et al.*, 2009). Economically, sustainability emphasizes efficient operations, life-cycle planning, and resource recovery approaches that provide value from wastewater, like biogas or nutrient-rich fertilizers (Zielińska & Bułkowska, 2024). Socially, sustainable wastewater management is important to equal access to safe sanitation, to public health, and to participation of residents in decision-making (Iribarnegaray & Seghezzo, 2012). To make wastewater systems resilient, adaptable, and beneficial to humans and the environment (Table 1.2), it is important to work with these three dimensions in a balanced holistic manner (Table 1.2).

1.3.2 Aligning wastewater systems with sustainable development goals

The UN SDGs (sustainable development goals) provide a global model for wastewater management (Delanka-Pedige *et al.*, 2021). Clean Water and Sanitation (SDG6), Responsible Consumption and Production (SDG 12), and Climate Action (SDG 13) are related directly to the planning and operation of wastewater systems (Kurniawan *et al.*, 2024). As mentioned by Obaideen *et al.* (2022), the appropriate management of wastewater should be responsive to these goals through the use of technologies and policies that are safe to treat, conducive to water use, energy and resource recovery, and reduce greenhouse gas emissions (Jain & Shah, 2025). By connecting operational decisions to broader sustainability objectives, wastewater systems can move from being purely regulatory compliance tools to drivers of environmental resilience, social equity, and economic efficiency (Srinivasan *et al.*, 2025).

Table 1.2 Sustainability dimensions in wastewater management.

Sustainability Dimension	Description	Example Practices	Expected Benefits
Environmental	Minimizing pollution and ecological impact	Nutrient recovery, advanced treatment, resource reuse	Reduced ecosystem contamination, improved water quality
Economic	Cost-effective operation and value creation	Energy recovery, biosolids valorization, operational efficiency	Reduced operational costs, revenue from recovered resources
Social	Public health, community engagement, and equity	Safe reuse of treated water, participatory programs	Improved health outcomes, increased public trust

1.4 CIRCULAR ECONOMY CONCEPTS APPLIED TO WASTEWATER SYSTEMS

1.4.1 From waste disposal to resource recovery

Traditional wastewater treatment has tended to involve collection, treatment, and safe disposal with wastewater being perceived as a burden rather than a resource (Kundu *et al.*, 2022). According to Li *et al.* (2023a), it is this perspective that the circular economy paradigm challenges, turning waste into goods and energy. By treating wastewater as a resource stream, and not a waste stream, systems can be designed to conserve energy, nutrients, and water at the same time reduce environmental pollution (Kundu *et al.*, 2022). This entails new technologies, coordinated planning, and economic incentives that encourage recovery over disposal. Li *et al.* (2023b) state that moving toward resource-based management is not only lessening the ecological footprint of wastewater systems, but has tangible benefits such as generation of renewable energy, fertilizer production, and safe drinking water reuse (Figure 1.3).

1.4.2 Nutrient, energy, and water reuse pathways

As per Kathi *et al.* (2023), wastewater contains large proportions of organic matter, nitrogen, phosphorus, and water that can be recovered and reused for productive activities. Renewal of nutrients can be achieved through struvite precipitation, biofilm treatment, or algae-based system, resulting in agricultural fertilizers without using synthetic inputs (Shepsko *et al.*, 2019). As outlined by Aluvihara *et al.* (2025), the recovery of energy, particularly from organic-rich sludge, can be enabled via anaerobic digestion, biogas capture, and co-generation, with the goal of energy self-sufficient treatment plants. Water

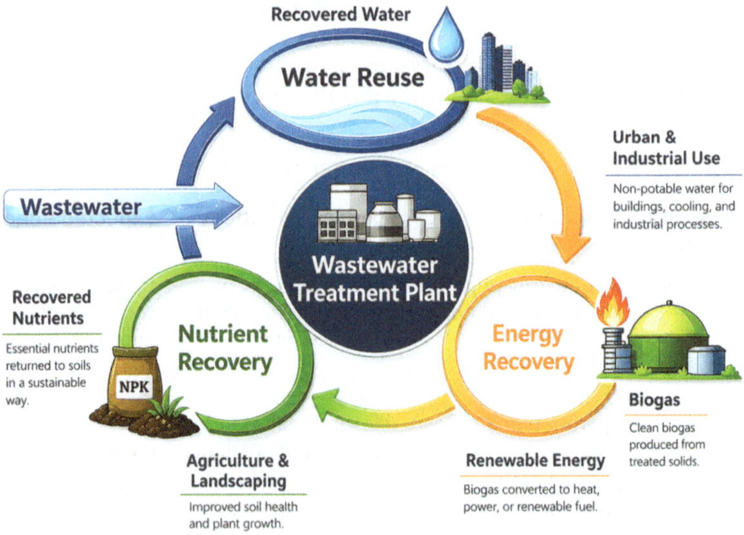

Figure 1.3 Circular economy in wastewater management.

reuse approaches can include irrigation, industrial cooling, and even indirect reuse of potable water after further treatment (Voulvoulis, 2018). These not only save resources but also foster sustainable agriculture, energy security, and resilience in cities, a core idea of the circular economy: extract the maximum value from waste streams (McCoy *et al.*, 2026).

1.4.3 Closing material and energy loops

Closing loops in wastewater systems require a continuous cycle of treatment, recovery, and reuse, rather than a one-way process from generation to disposal (Van Fan *et al.*, 2019). Circular strategies aim to reduce resource losses, greenhouse gas emissions, and create synergies within sectors. For example, reclaimed nutrients can fertilize nearby farms, biogas can power plants or local communities, and treated water can supply urban green space or industrial processes (Wang *et al.*, 2022). With the inclusion of these loops in planning and operation, wastewater management is a major generator of sustainability in urban and agricultural economies (Capodaglio, 2017). In addition, digital monitoring, intelligent sensors, and predictive analytics can help improve loop efficiency, with optimal recovery and minimal losses (Lee *et al.*, 2020). In the long run, closing material and energy loops reorients wastewater systems away from environmental burdens into pillars of circular and resilient urban infrastructure.

1.5 EVOLUTION OF WASTEWATER TREATMENT SYSTEMS

1.5.1 Conventional centralized treatment models

These types of wastewater treatment systems have traditionally focused on centrally located treatment plants that collect waste from cities and business areas, run through large sewer lines and discharge treated wastes to rivers, lakes, or seashores (Capodaglio, 2017). These based on activated sludge systems, trickling filters, oxidation ponds, or stabilization tanks were designed to extract only basic pollutants and reduce pathogens at scale (Shamshad & Rehman, 2024). Centralized plants have been the backbone of modern urban sanitation for decades because they offer economies of scale, systematic monitoring, and controlled operations. But according to Silva (2023), while centralized models are effective in reducing pollutants, they have at least some shortcomings. Chernicharo *et al.* (2015) state that they require big investments in infrastructure, such as long pipelines, pump stations, and large treatment facilities, but may not be a feasible option for increasingly dense cities or resource-poor regions. Operational challenges such as energy demand, sludge management, and vulnerability to system failures also constrain their flexibility (Silva, 2023). Moreover, conventional plants are typically designed with a linear approach, emphasizing pollutant removal over resource recovery, limiting their contribution to energy, nutrient, and water reuse (Kehrein *et al.*, 2020). These constraints have prompted researchers and policymakers to explore more adaptive, resilient, and resource-efficient approaches (Ventura *et al.*, 2024).

1.5.2 Emergence of decentralized and hybrid systems

According to Sha *et al.* (2024), decentralized and hybrid wastewater treatment approaches have come to replace centralized systems. They further note that decentralized systems discharge wastewater closer to the source in the community, housing unit, or industrial area, and reduce the use of long-distance sewer lines and the possibility of untreated overflows or system overload. They are often modular and scalable, which means they can quickly be utilized in urban areas, peri-urban settlements, or areas where populations are unevenly distributed (Table 1.3).

Hybrid systems combine centralized treatment plants with smaller, localized treatment units, creating a multifaceted network able to cope with varying flows and pollutant loads (Liu *et al.*, 2023). Buller *et al.* (2021) state that industrial wastewaters, for example, can be transported to the local landfill and mixed with municipal waste for treatment further in a central facility. Decentralized and hybrid systems are also an ideal location for incorporating new technologies such as membrane bioreactors, constructed wetlands, anaerobic digesters, and bioelectrochemical systems that can be used simultaneously for treatment, energy recovery, and nutrient reclamation (Yu *et al.*, 2024). Through these integrations, contemporary wastewater systems offer resilience but also actively contribute toward sustainable water management and circular economy (Liu *et al.*, 2023).

1.5.3 Drivers of technological transformation

Multiple related factors have been involved in the transition from conventional to advanced wastewater treatment systems (Sangamnere *et al.*, 2023). Urbanization and population growth continue to increase the volume of wastewater, but industrial diversification introduces a variety of chemical and

Table 1.3 Comparative overview of wastewater treatment models.

Model Type	Key Features	Advantages	Limitations	Example Applications
Centralized	Large-scale plants, centralized collection network	Economies of scale; uniform treatment quality	High infrastructure cost; inflexible; vulnerable to failures	Urban municipal wastewater treatment
Decentralized	Small-scale, localized treatment units	Flexible; reduces transport; suitable for remote or peri-urban areas	Lower efficiency for large flows; requires local management	Rural communities, small towns, industrial clusters
Hybrid	Combination of centralized and decentralized systems	Balances efficiency and flexibility; supports phased expansion	Complex design and operation; higher planning requirements	Growing cities with mixed urban–rural zones

biological contaminants that threaten traditional treatment approaches (Tella *et al.*, 2024). Climate change, water scarcity, and extreme weather events add to system vulnerability, and require adaptive and robust systems that will work in varying environments (Thornton *et al.*, 2014). Economic and social pressures also have an important role. There is increasing emphasis on reducing operational costs, optimizing energy consumption, and recovering water, nutrients, and energy from wastewater streams (Li *et al.*, 2023a). Regulatory and policymaking increasingly favor sustainable management, environmental protection, and the application of innovative treatment practices (Metz & Ingold, 2014). Plus, digitalization, monitoring through Internet of Things (IoT) and predictive analytics provide real-time performance monitoring, fault-finding, and data-driven optimization that increases efficiency and resilience (Wagh & Agashe, 2023). Together, these factors have transformed the guise of wastewater management from linear, disposal-orientated systems to adaptive resource-orientated and integrated solutions (Zhang & Liu, 2022). Zhang (2025) states that modern treatment systems are now designed to not just protect human health and the environment, but to recover valuable resources, reduce carbon emissions, and contribute to broad sustainability goals.

1.6 SMART TECHNOLOGIES AND SYSTEM INTEGRATION

1.6.1 Role of sensing, automation, and real-time monitoring

Smart technologies are increasingly being used in the context of modern wastewater treatment to improve performance, reliability, and environmental performance (Alprol *et al.*, 2024). Sensors and automation systems are critical to continuously monitoring a number of key parameters such as flow rate, pH, dissolved oxygen, nutrients, and pollutants (Zainurin *et al.*, 2022). Real-time monitoring helps operators detect anomalies, prevent system overflow, and respond proactive to operational problems. Pumps, valves, and aeration systems can be controlled by automation so treatment plants can continuously improve processes, thereby improving efficiency and lowering energy consumption (Drewnowski *et al.*, 2019). As per De La Hoz *et al.* (2026), sensory data and automated control are the foundational components of intelligent wastewater systems, replacing conventional infrastructure with adaptive management practices (Figure 1.4).

1.6.2 Data-driven decision-making in wastewater operations

According to Dikmen *et al.* (2025), the combination of data analysis, prediction modeling, and machine learning has turned wastewater treatment from reactive to proactive decision-making. Forecasting influent characteristics, optimal treatment cycles, equipment maintenance needs, and resource recovery are all possible from historical and real-time data (Zhou *et al.*, 2022). Predictive models, for example, can anticipate high organic loads or nutrient content to enable pre-emptive change in aeration or sludge handling (Xiong *et al.*, 2025). Data-driven approaches are also appropriate for performance benchmarking, regulatory compliance, and sustainability reporting that helps managers make informed decisions based on environmental, economic, and social goals

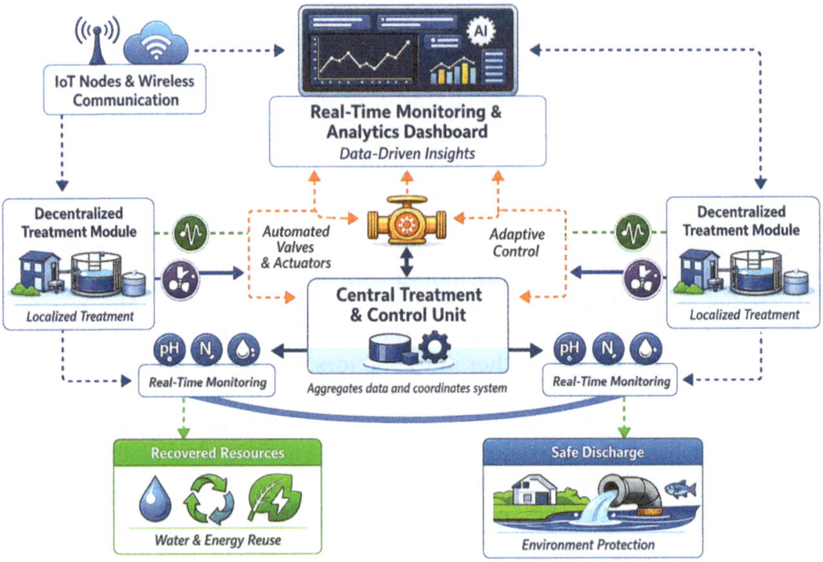

Figure 1.4 Smart wastewater systems integration.

(Dikmen *et al.*, 2025). Mao *et al.* (2024) state that analytics have enabled wastewater systems to become more efficient, resilient, and consistent with circular economy principles.

1.6.3 Building adaptive and resilient infrastructures

De La Hoz *et al.* (2026) state that smart technologies and integrated system design result in adaptive and resilient wastewater infrastructures. Resilience translates into the ability to maintain service in otherwise unpredictable conditions, to recover quickly from disruptions and to take advantage of future technological advancement (Li *et al.*, 2023b). The adapted infrastructures include modular systems, decentralized treatment units, and a mix of operational and digital monitoring tools designed to cope with rapidly changing flows, extreme weather conditions, and incoming contaminants (Sha *et al.*, 2024). According to Dikmen *et al.* (2025), real-time data feedback loops allow operators to improve energy use, resource recovery, and minimize environmental impacts. Finally, intelligent and integrated systems turn wastewater networks into self-governing dynamic units able to respond to urban sustainability, public health, and environmental management (Mao *et al.*, 2024).

1.7 POSITIONING WASTEWATER MANAGEMENT WITHIN URBAN AND ENVIRONMENTAL SYSTEMS

The world of wastewater today, therefore, is far beyond the control of pollution, and is a composite energy, agriculture, and water use (Tella *et al.*, 2024).

Table 1.4 Wastewater management and smart city linkages.

Smart City Component	Role of Wastewater System	Technology Used	Benefits
Urban water supply	Reuse treated wastewater for irrigation or industrial use	Sensors, real-time monitoring	Reduces freshwater demand, ensures sustainable supply
Energy management	Recover energy from sludge and biogas production	Anaerobic digesters, smart meters	Renewable energy generation, cost reduction
Public health & safety	Detect contamination and prevent outbreaks	IoT sensors, predictive analytics	Timely alerts, improved public health
Urban planning & infrastructure	Integrate wastewater systems into smart city design	GIS mapping, digital twins	Optimized infrastructure, resource efficiency

As mentioned by Opperman *et al.* (2022), treated waste can be a viable source of water for irrigation, a tool for sustainable farming and a source to reduce the demand for freshwater. Processed organic sludge and biosolids can be converted into biogas or biofertilizers which provide renewable energy and nutrient-rich soil amendments. The interconnections between these paths allow cities to recover surplus resources, reduce carbon footprint and close material loops that comply with circular economy principles (McCoy *et al.*, 2026). By harnessing these connections, wastewater systems transform from being a burden into a resource, and so foster sustainability in the environment and economic recovery (Kathi *et al.*, 2023).

Wastewater systems are becoming increasingly identified as a vital infrastructure to sustainable cities in the vision of smart cities (Delanka-Pedige *et al.*, 2020). This combination of IoT sensors, real-time monitoring, predictive analytics, and automated control enables cities to better regulate treatment operations, identify system failures, and ensure consistent services (De La Hoz *et al.*, 2026). Smart wastewater networks can also function with other urban utilities, such as energy grids or water distribution systems, bringing synergistic efficiencies (Sha *et al.*, 2024). Biogas, for example, provides energy for treatment or the local grid, and recycled water helps to grow cities or feed industries. Through bringing wastewater treatment into the digital and practical fabric of smart cities, municipalities can create resilient, adapted and resource-efficient urban ecosystems that respond to the growing population need while also protecting environmental and public health (Table 1.4).

1.8 SCOPE, STRUCTURE, AND ORGANIZATION OF THE BOOK

This book has been selected to provide a comprehensive overview of the most advanced wastewater management concepts, technologies, and applications. Each chapter considers a specific, but closely related, questioning of wastewater systems, from sustainable design, circular economy, and smart technologies to energy recovery, data-driven operations, and case studies. These chapters are designed to move from concept to practice so that readers can appreciate both the theoretical and practical aspects of modern wastewater management. Through the book's highlighting of emerging trends, policy and resource recovery agendas, it provides a holistic view of the sector that bridges the gap between research, practice, and policy.

Wastewater management is a natural and complex process, involving environmental engineering, civil engineering, data science, microbiology, energy management, and policy studies (Matheri *et al.*, 2022). In this spirit the book draws on data across many domains, offering concepts accessible to all readers without impairing technical understanding. Intended readership includes advanced undergraduates, postgraduate students, researchers, environmental professionals, policymakers, and engineers who need both conceptual clarity and applied knowledge. The book provides rigorous scientific analysis, contextual context, and practical examples that provide readers with tools to design, evaluate, and implement sustainable and smart wastewater management solutions that enables innovation and informed decision making both in the academic, industrial, and municipal settings.

1.9 CONCLUSION

This chapter has highlighted the transformation from a simple filtration task to an integral part of sustainable urban and environmental planning. Traditional centralized treatment systems laid the foundation; but, their limitations in flexibility, resilience, and resource recovery have meant the construction of decentralized and hybrid treatment systems that can more effectively respond to urban growth, industrial development and climate change. Sustainability and circular economy principles lie at the heart of this shift. Wastewater is increasingly recognized as a resource of the future for the recovery of water, nutrients, and energy. These recovered resources could support agriculture, generate energy from renewable sources and contribute to broader urban and environmental resilience. As automation, sensors, and real-time monitoring are integrated with data-driven decisions, wastewater treatment plants can be made more efficient, more proactive in addressing problems and more responsive to changing situations. This chapter sets the stage for the rest of the book by situating wastewater management within an integrated engineering, environmental science, policy, and digital technology system. It demonstrates that modern wastewater systems are not just technical infrastructures, but strategic assets for sustainable development. These insights serve as a guide to other promising technologies, energy recovery, and applications in the later chapters.

REFERENCES

Alprol A. E., Mansour A. T., Ibrahim M. E. E. and Ashour M. (2024). Artificial intelligence technologies revolutionizing wastewater treatment: current trends and future prospective. *Water*, **16**(2), 314, https://doi.org/10.3390/w16020314

Aluvihara S., Alam S., Omar M., Hilonga A., Zaryab A. and Sadeg S. (2025). Eco-friendly and cost-effective water treatment and wastewater treatment technologies: a review. *American Journal of Water Science and Engineering*, **11**(3), 74–85, https://doi.org/10.11648/j.ajwse.20251103.13

Batuwita S., Udugampala S. and Edirisinghe U. (2022). Factors Affecting Habitat Selection of Freshwater Fishes of Sri Lanka: A Review. Apple Academic Press, New York, pp. 201–262, https://doi.org/10.1201/9781003277125-11

Buller L., Sganzerla W., Berni M., Brignoli S. and Forster-Carneiro T. (2021). Design and techno-economic analysis of a hybrid system for energy supply in a wastewater treatment plant: a decentralized energy strategy. *Journal of Environmental Management*, **305**, 114389, https://doi.org/10.1016/j.jenvman.2021.114389

Capodaglio A. (2017). Integrated, decentralized wastewater management for resource recovery in rural and peri-urban areas. *Resources*, **6**(2), 22, https://doi.org/10.3390/resources6020022

Chernicharo C. A. L., Van Lier J. B., Noyola A. and Ribeiro T. B. (2015). Anaerobic sewage treatment: state of the art, constraints and challenges. *Reviews in Environmental Science and Bio/Technology*, **14**(4), 649–679, https://doi.org/10.1007/s11157-015-9377-3

Das A. and Mishra S. (2025). Smart and sustainable water management: policy, technology, and innovation pathways for resource resilience. *Green Technology Resilience and Sustainability*, **5**(1), 1–26, https://doi.org/10.1007/s44173-025-00022-8

De La Hoz J. M., Islam M. D. S., Bappy M. M. and Hayes M. P. (2026). AI-enabled modeling for smart rural wastewater treatment systems: current practices and remaining gaps. *Applied Water Science*, **16**(2), 40, https://doi.org/10.1007/s13201-025-02698-6

Delanka-Pedige H. M. K., Munasinghe-Arachchige S. P., Abeysiriwardana-Arachchige I. S. A. and Nirmalakhandan N. (2020). Wastewater infrastructure for sustainable cities: assessment based on UN sustainable development goals (SDGs). *International Journal of Sustainable Development & World Ecology*, **28**(3), 203–209, https://doi.org/10.1080/13504509.2020.1795006

Delanka-Pedige H. M. K., Munasinghe-Arachchige S. P., Abeysiriwardana-Arachchige I. S. A. and Nirmalakhandan N. (2021). Evaluating wastewater treatment infrastructure systems based on UN sustainable development goals and targets. *Journal of Cleaner Production*, **298**, 126795, https://doi.org/10.1016/j.jclepro.2021.126795

Dikmen F., Demir A., Özkaya B., Raza M. O., Rasheed J., Asuroglu T. and Alsubai S. (2025). AI-driven wastewater management through comparative analysis of feature selection techniques and predictive models. *Scientific Reports*, **15**(1), 25347, https://doi.org/10.1038/s41598-025-07124-0

Drewnowski J., Remiszewska-Skwarek A., Duda S. and Łagód G. (2019). Aeration process in bioreactors as the main energy consumer in a wastewater treatment plant. Review of solutions and methods of process optimization. *Processes*, **7**(5), 311, https://doi.org/10.3390/pr7050311

Dutta D., Arya S. and Kumar S. (2021). Industrial wastewater treatment: current trends, bottlenecks, and best practices. *Chemosphere*, **285**, 131245, https://doi.org/10.1016/j.chemosphere.2021.131245

Garg M. C., Kumari S. and Agarwal S. (2024). The integration of artificial intelligence in advanced wastewater treatment systems. In: Garg M. C. (ed.) *The AI Cleanse: Transforming Wastewater Treatment Through Artificial Intelligence,* Springer Water, Springer, Cham, pp. 1–27, https://doi.org/10.1007/978-3-031-67237-8_1

Guest J. S., Skerlos S. J., Barnard J. L., Beck M. B., Daigger G. T., Hilger H., Jackson S. J., Karvazy K., Kelly L., Macpherson L., Mihelcic J. R., Pramanik A., Raskin L., Van Loosdrecht M. C. M., Yeh D. and Love N. G. (2009). A new planning and design

paradigm to achieve sustainable resource recovery from wastewater. *Environmental Science & Technology*, **43**(16), 6126–6130, https://doi.org/10.1021/es9010515
Hobbie S. E. and Grimm N. B. (2020). Nature-based approaches to managing climate change impacts in cities. *Philosophical Transactions of the Royal Society B Biological Sciences*, **375**(1794), 20190124, https://doi.org/10.1098/rstb.2019.0124
Hyde-Smith L., Zhan Z., Roelich K., Mdee A. and Evans B. (2022). Climate change impacts on urban sanitation: a systematic review and failure mode analysis. *Environmental Science & Technology*, **56**(9), 5306–5321, https://doi.org/10.1021/acs.est.1c07424
Iribarnegaray M. and Seghezzo L. (2012). Governance, sustainability and decision making in water and sanitation management systems. *Sustainability*, **4**(11), 2922–2945, https://doi.org/10.3390/su4112922
Jain H. and Shah M. P. (2025). Microbial Insights Into Wastewater Treatment and Environmental Sustainability. Bentham Science Publishers Singapore PTE. LTD, Singapore.
Kathi S., Singh S., Yadav R., Singh A. N. and Mahmoud A. E. D. (2023). Wastewater and sludge valorisation: a novel approach for treatment and resource recovery to achieve circular economy concept. *Frontiers in Chemical Engineering*, **5**, 1129783, https://doi.org/10.3389/fceng.2023.1129783
Kehrein P., Van Loosdrecht M., Osseweijer P., Garfí M., Dewulf J. and Posada J. (2020). A critical review of resource recovery from municipal wastewater treatment plants–market supply potentials, technologies and bottlenecks. *Environmental Science Water Research & Technology*, **6**(4), 877–910, https://doi.org/10.1039/c9ew00905a
Kundu D., Dutta D., Samanta P., Dey S., Sherpa K. C., Kumar S. and Dubey B. K. (2022). Valorization of wastewater: a paradigm shift towards circular bioeconomy and sustainability. *The Science of the Total Environment*, **848**, 157709, https://doi.org/10.1016/j.scitotenv.2022.157709
Kurniawan T. A., Mohyuddin A., Casila J. C. C., Sarangi P. K., Al-Hazmi H., Wibisono Y., Kusworo T. D., Khan M. M. H. and Haddout S. (2024). Digitalization for sustainable wastewater treatment: a way forward for promoting the UN SDG#6 'clean water and sanitation' towards carbon neutrality goals. *Discover Water*, **4**(1), 71, https://doi.org/10.1007/s43832-024-00134-5
Lee J., Ni J., Singh J., Jiang B., Azamfar M. and Feng J. (2020). Intelligent maintenance systems and predictive manufacturing. *Journal of Manufacturing Science and Engineering*, **142**(11), 110805, https://doi.org/10.1115/1.4047856
Lehmann S. (2017). Implementing the urban nexus approach for improved resource-efficiency of developing cities in Southeast-Asia. *City Culture and Society*, **13**, 46–56, https://doi.org/10.1016/j.ccs.2017.10.003
Li J., Li X., Liu H., Gao L., Wang W., Wang Z., Zhou T. and Wang Q. (2023a). Climate change impacts on wastewater infrastructure: a systematic review and typological adaptation strategy. *Water Research*, **242**, 120282, https://doi.org/10.1016/j.watres.2023.120282
Li X., Van Der Bruggen B., Matsuyama H., Lin Y. and Zheng J. (2023b). Resource recovery and recycling from water streams: advanced membrane technologies and case studies. *ACS ES&T Water*, **3**(7), 1699–1701, https://doi.org/10.1021/acsestwater.3c00315
Liu L., Zhou X., Dueñas-Osorio L., Stadler L. and Li Q. (2023). Hybrid wastewater treatment and reuse enhances urban water system resilience to disruptive incidents. *Nature Water*, **1**(12), 1048–1058, https://doi.org/10.1038/s44221-023-00166-6
Makanda K., Nzama S. and Kanyerere T. (2022). Assessing the role of water resources protection practice for sustainable water resources management: a review. *Water*, **14**(19), 3153, https://doi.org/10.3390/w14193153
Mao Z., Li X., Zhang X., Li D., Lu J., Li J. and Zheng F. (2024). Optimization of effluent quality and energy consumption of aeration process in wastewater treatment plants

using artificial intelligence. *Journal of Water Process Engineering*, **63**, 105384, https://doi.org/10.1016/j.jwpe.2024.105384

Matheri A. N., Mohamed B., Ntuli F., Nabadda E. and Ngila J. C. (2022). Sustainable circularity and intelligent data-driven operations and control of the wastewater treatment plant. *Physics and Chemistry of the Earth Parts A/B/C*, **126**, 103152, https://doi.org/10.1016/j.pce.2022.103152

McCoy D., Hardy H. and Townsend R. (2026). Circular Economy in Wastewater Management. ResearchGate, https://www.researchgate.net/publication/399959067_Circular_Economy_in_Wastewater_Management

Metz F. and Ingold K. (2014). Sustainable wastewater management: is it possible to regulate micropollution in the future by learning from the past? A policy analysis. *Sustainability*, **6**(4), 1992–2012, https://doi.org/10.3390/su6041992

Muga H. E. and Mihelcic J. R. (2007). Sustainability of wastewater treatment technologies. *Journal of Environmental Management*, **88**(3), 437–447, https://doi.org/10.1016/j.jenvman.2007.03.008

Obaideen K., Shehata N., Sayed E. T., Abdelkareem M. A., Mahmoud M. S. and Olabi A. (2022). The role of wastewater treatment in achieving sustainable development goals (SDGs) and sustainability guideline. *Energy Nexus*, **7**, 100112, https://doi.org/10.1016/j.nexus.2022.100112

Öberg G., Metson G. S., Kuwayama Y. and Steven A. C. (2020). Conventional Sewer Systems are too Time-Consuming, Costly and Inflexible to Meet the Challenges of the 21st Century. *Sustainability*, **12**(16), 6518, https://doi.org/10.3390/su12166518

Opperman J. J., Camargo R. R., Laporte-Bisquit A., Zarfl C. and Morgan A. J. (2022). Using the WWF water risk filter to screen existing and projected hydropower projects for climate and biodiversity risks. *Water*, **14**(5), 721, https://doi.org/10.3390/w14050721

Pastor J. and Hernández A. (2011). Heavy metals, salts and organic residues in old solid urban waste landfills and surface waters in their discharge areas: determinants for restoring their impact. *Journal of Environmental Management*, **95**, S42–S49, https://doi.org/10.1016/j.jenvman.2011.06.048

Rafique M., Hajra S., Tahir M. B., Gillani S. S. A. and Irshad M. (2022). A review on sources of heavy metals, their toxicity and removal technique using physico-chemical processes from wastewater. *Environmental Science and Pollution Research*, **29**(11), 16772–16781, https://doi.org/10.1007/s11356-022-18638-9

Ravina M., Galletta S., Dagbetin A., Kamaleldin O. A. H., Mng'ombe M., Mnyenyembe L., Shanko A. and Zanetti M. (2021). Urban wastewater treatment in African countries: evidence from the hydroaid initiative. *Sustainability*, **13**(22), 12828, https://doi.org/10.3390/su132212828

Rezai B. and Allahkarami E. (2021). Wastewater Treatment Processes–Techniques, Technologies, Challenges Faced, and Alternative Solutions. In Soft computing techniques in solid waste and wastewater management, Elsevier, pp. 35–53, https://doi.org/10.1016/b978-0-12-824463-0.00004-5

Sangamnere R., Misra T., Bherwani H., Kapley A. and Kumar R. (2023). A critical review of conventional and emerging wastewater treatment technologies. *Sustainable Water Resources Management*, **9**(2), 58, https://doi.org/10.1007/s40899-023-00829-y

Sha C., Shen S., Zhang J., Zhou C., Lu X. and Zhang H. (2024). A review of strategies and technologies for sustainable decentralized wastewater treatment. *Water*, **16**(20), 3003, https://doi.org/10.3390/w16203003

Shah M. P. (ed.) (2024). Developments in Wastewater Treatment Research and Processes: Applied Technologies for Clean-up of Environmental Contaminants. Elsevier, Amsterdam, Netherlands.

Shah M. P. and Rodriguez-Couto S. (eds) (2019). Microbial Wastewater Treatment. Elsevier, Amsterdam, Netherlands.

Shamshad J. and Rehman R. U. (2024). Innovative approaches to sustainable wastewater treatment: a comprehensive exploration of conventional and emerging technologies. *Environmental Science Advances*, **4**(2), 189–222, https://doi.org/10.1039/d4va00136b

Shepsko C. S., Dong H. and SenGupta A. K. (2019). Treated municipal wastewater reuse: a holistic approach using hybrid ion exchange (HIX) with concurrent nutrient recovery and CO_2 sequestration. *ACS Sustainable Chemistry & Engineering*, **7**(10), 9671–9679, https://doi.org/10.1021/acssuschemeng.9b01357

Silva J. A. (2023). Wastewater treatment and reuse for sustainable water resources management: a systematic literature review. *Sustainability*, **15**(14), 10940, https://doi.org/10.3390/su151410940

Srinivasan L., Sowjanya S., Chakraborty A., Amrit R., Nurgat M. Z., Dutta P., Das P., Das A., Sivarajan S. and Wagh M. (2025). Economic, Environmental, and Social Optimization of Wastewater Management in the Context of Circular Economy. In Water Use Efficiency, Sustainability and The Circular Economy, Elsevier, Amsterdam, Netherlands, pp. 203–230, https://doi.org/10.1016/b978-0-443-26749-9.00011-9

Stec J., Kosikowska U., Mendrycka M., Stępień-Pyśniak D., Niedźwiedzka-Rystwej P., Bębnowska D., Hrynkiewicz R., Ziętara-Wysocka J. and Grywalska E. (2022). Opportunistic pathogens of recreational waters with emphasis on antimicrobial resistance–A possible subject of human health concern. *International Journal of Environmental Research and Public Health*, **19**(12), 7308, https://doi.org/10.3390/ijerph19127308

Sweetapple C., Fu G., Farmani R. and Butler D. (2018). Exploring wastewater system performance under future threats: does enhancing resilience increase sustainability? *Water Research*, **149**, 448–459, https://doi.org/10.1016/j.watres.2018.11.025

Tella T. A., Festus B., Olaoluwa T. D. and Oladapo A. S. (2024). Water and Wastewater Treatment in Developed and Developing Countries: Present Experience and Future Plans. In Smart nanomaterials for environmental applications, Elsevier, Amsterdam, Netherlands, pp. 351–385, https://doi.org/10.1016/b978-0-443-21794-4.00030-2

Thornton P. K., Ericksen P. J., Herrero M. and Challinor A. J. (2014). Climate variability and vulnerability to climate change: a review. *Global Change Biology*, **20**(11), 3313–3328, https://doi.org/10.1111/gcb.12581

Van Fan Y., Lee C. T., Lim J. S., Klemeš J. J. and Le P. T. K. (2019). Cross-disciplinary approaches towards smart, resilient and sustainable circular economy. *Journal of Cleaner Production*, **232**, 1482–1491, https://doi.org/10.1016/j.jclepro.2019.05.266

Ventura J. S., Tulipan J. U., Banawa A., Umali K. D. and Villanueva J. A. L. (2024). Advancements and challenges in decentralized wastewater treatment: a comprehensive review. *Desalination and Water Treatment*, **320**, 100830, https://doi.org/10.1016/j.dwt.2024.100830

Voulvoulis N. (2018). Water reuse from a circular economy perspective and potential risks from an unregulated approach. *Current Opinion in Environmental Science & Health*, **2**, 32–45, https://doi.org/10.1016/j.coesh.2018.01.005

Vs S., Mollinga P. P. and Bogardi J. J. (2011). Global change, wastewater and health in fast growing economies. *Current Opinion in Environmental Sustainability*, **3**(6), 461–466, https://doi.org/10.1016/j.cosust.2011.10.009

Wagh N. and Agashe S. D. (2023). Monitoring exhaust air temperature and detecting faults during milk spray drying using data-driven framework. *Drying Technology*, **41**(12), 1991–2011, https://doi.org/10.1080/07373937.2023.2213312

Wang J., Sun Y., Xia K., Deines A., Cooper R., Pallansch K. and Wang Z. (2022). Pivotal role of municipal wastewater resource recovery facilities in urban agriculture: a review. *Water Environment Research*, **94**(6), e10743, https://doi.org/10.1002/wer.10743

Wu X., Nawaz S., Li Y. and Zhang H. (2024). Environmental health hazards of untreated livestock wastewater: potential risks and future perspectives. *Environmental Science and Pollution Research*, **31**(17), 24745–24767, https://doi.org/10.1007/s11356-024-32853-6

Xiong Z., Cheng L., Yunus A. I., Jian D., Zhang X., Igou T., Tong S. and Chen Y. (2025). Machine learning-assisted expert control in wastewater aeration treatment process application. *ACS ES&T Engineering*, **5**(11), 3140–3152, https://doi.org/10.1021/acsestengg.5c00493

Yu W., Liu H., Lin H., Lee M. and Hou C. (2024). Synergizing MBR and MCDI systems as a sustainable solution for decentralized wastewater reclamation and reuse. *Sustainable Environment Research*, **34**(1), 12, https://doi.org/10.1186/s42834-024-00217-x

Zainurin S. N., Ismail W. Z. W., Mahamud S. N. I., Ismail I., Jamaludin J., Ariffin K. N. Z. and Kamil W. M. W. A. (2022). Advancements in monitoring water quality based on various sensing methods: a systematic review. *International Journal of Environmental Research and Public Health*, **19**(21), 14080, https://doi.org/10.3390/ijerph192114080

Zhang T. (2025). Sustainable wastewater treatment and the circular economy. *Water*, **17**(3), 335, https://doi.org/10.3390/w17030335

Zhang, X. and Liu, Y. (2022). Circular economy is game-changing municipal wastewater treatment technology towards energy and carbon neutrality. *Chemical Engineering Journal*, **429**, 132114, https://doi.org/10.1016/j.cej.2021.132114

Zhou P., Li Z., Snowling S., Goel R. and Zhang Q. (2022). Multi-step ahead prediction of hourly influent characteristics for wastewater treatment plants: a case study from North America. *Environmental Monitoring and Assessment*, **194**(5), 389, https://doi.org/10.1007/s10661-022-09957-y

Zielińska M. and Bułkowska K. (2024). Sustainable management and advanced nutrient recovery from biogas energy sector effluents. *Energies*, **17**(15), 3705, https://doi.org/10.3390/en17153705

doi: 10.2166/9781789065404_0019

Chapter 2

Technologies for smart wastewater collection

Gourav Sabharwal[1]*, *K. Vaiyapuri*[1], *Aparna Mohan V.*[2], *Mummasani Asritha*[1], *S. Selvakumar*[3], *Vivek Sharma*[4], *R. Jagadeeswaran*[5] and *M. Raju*[1]

[1]*Department of Agronomy, Tamil Nadu Agricultural University, Coimbatore, India*
[2]*ICAR- Central Tuber Crop Research Institute, Sreekariyam, Thiruvananthapuram, Kerala, India*
[3]*Center for Water and Geospatial Studies, Tamil Nadu Agricultural University, Coimbatore, India*
[4]*Agricultural and Biological Engineering Department, University of Florida, Gainesville, FL 32611, USA*
[5]*Department of RS&GIS, Tamil Nadu Agricultural University, Coimbatore, India*
**Correspondence mail: sabharwal308@gmail.com*

ABSTRACT

Smart wastewater collection has evolved into a robust system for extracting and refining pollutants from wastewater which can then be reused with minimal environmental impact. New technologies use sensors, the Internet of Things (IoT), and information and communication technologies to provide real-time monitoring and enable smart wastewater management systems. Advanced techniques based on blockchain technology enable efficient utilization of water and energy in agriculture and water treatment plants. Water is a crucial resource on the planet for achieving the sustainable development goals 2030, and wastewater collection is a crucial element of that drive. Previously adopted traditional wastewater treatment techniques, such as thermal treatment adsorption, coagulation, and electrochemical degradation can have drawbacks such as toxic effects, and high costs. This chapter focuses on new smart remote sensing and IoT technologies, advances in sensors and data transfer technologies, and data-driven smart waste collection processes that can overcome old inefficient collection methods. Good quality water availability is important for maintaining life quality as well as preventing waterborne diseases. These state-of-the-art technologies lead to improved wastewater collection which contributes to reducing water scarcity worldwide, but their implementation faces some significant challenges relating to wireless data transfer, battery lifetime and IoT infrastructure. Therefore, the development of affordable solutions and robust infrastructure is essential for the adoption of smart wastewater technologies worldwide.

Keywords: Sensors, Internet of Things, sustainable development goals, information and communication technology

2.1 INTRODUCTION

Depletion of water resources and their contamination have emerged as two global dilemmas of the twenty-first century. According to the United Nations, about 2.3 billion people dwell in water-stressed regions and this figure is anticipated to grow further due to escalating demands associated with climate change, urbanization, and population growth increase demand (UN-Water, 2021). Rapid industrialization along with intensive agricultural practices has exacerbated freshwater scarcity resulting in depletion of aquifers, river contamination, and ecological degradation (UNESCO, 2019). Climate change has exacerbated these challenges and disrupted water supply and quality by increasing the frequency and severity of droughts and floods worldwide (Boretti & Rosa, 2019).

Furthermore, rapid urbanization has led to the generation of large volumes of untreated or insufficiently treated sewage. Worldwide, about 80% of sewage water is discharged untreated into the ecosystem posing significant threats to human health and aquatic biodiversity. This contaminated water contains many pollutants such as microplastics and pharmaceuticals, as well as heavy metals, nutrients, and pathogens which challenge conventional treatments and eco-friendly water reusage (Figure 2.1).

Sustainable and effective management of wastewater has become a key pillar of global water sustainability. The United Nations Sustainable Development Goal 6.3 aims to enhance water quality by reducing untreated wastewater and substantially increasing recycling and safe reuse by 2030 (UN-Water, 2021). Now, with technological advancements and the scarcity of freshwater, wastewater is considered a valuable resource rather than a waste product. It has

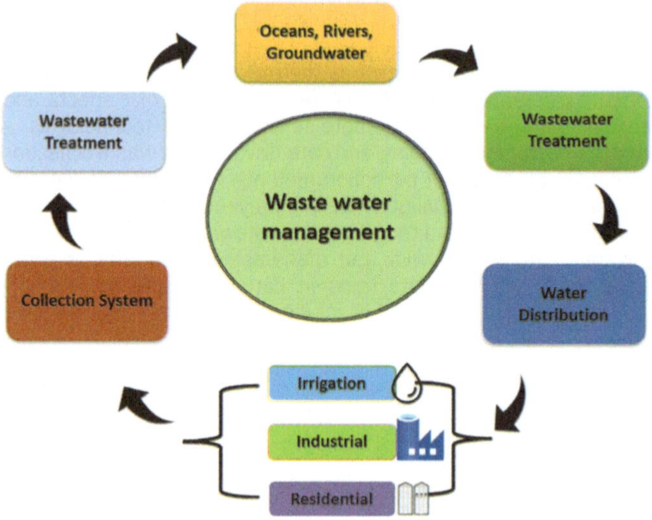

Figure 2.1 Wastewater management cycle.

the potential to provide clean water, recover nutrients, and generate renewable energy. Advanced treatment facilities reclaim wastewater for irrigation, industrial cooling, and potable reuse in water-scarce regions such as Singapore and California (Angelakis *et al.*, 2018).

Apart from augmenting water supply, efficient wastewater management reduces pollution, alleviates greenhouse gas emissions from untreated matter and improves human health by preventing the spread of waterborne diseases. Centralized treatment plants, integrated with decentralized solutions in peri-urban and rural areas, ensure that wastewater management provides sustainability along with resilience and equity (UNESCO, 2019). Nevertheless, traditional wastewater management encounters numerous barriers, including aging infrastructure, high operational expenses and limited real-time monitoring capabilities.

2.2 CONVENTIONAL WASTEWATER TREATMENT TECHNIQUES

Wastewater treatment is important in protecting the environment and human health particularly with increased urbanization and industrial activity. Despite the increased usage of advanced techniques such as membrane filtration and advanced oxidation, conventional/traditional techniques still remain the backbone of global treatment systems. This includes thermal treatment, adsorption, and electrochemical degradation which are widely used.

Thermal treatment methods include incineration, pyrolysis, evaporation, and wet air oxidation. These techniques employ heat in order to breakdown, decompose, or oxidize organic pollutants and pathogens. Indicatively, wet air oxidation has been utilized to treat very highly concentrated effluents with COD levels exceeding 100 000 mg/L, where refractory compounds are broken down significantly (Yavuz, 2007).

Thermal processes are advantageous due to their ability to fully eliminate pathogens and recalcitrant pollutants while significantly reducing sludge volume (Metcalf & Eddy/AECOM, 2014). But their applicability is limited by high energy consumption and capital costs. In addition, incomplete oxidation generates toxic and hazardous byproducts comprising of dioxins and furans, which pose environmental and human health risks (Pera-Titus *et al.*, 2005). Consequently, thermal methods are primarily restricted to specialized industrial effluents, such as those from the petrochemical and pharmaceutical sectors.

Adsorption eliminates organic and inorganic contaminants by transferring solutes onto the surface of porous solids such as activated carbon, zeolites, or biochar. Activated carbon possesses a high surface area and strong pollutant affinity which makes it a widely used absorbent (Mohan & Pittman, 2006). This approach is easy, cost-effective and efficient, particularly in removing trace pollutants like pesticides, dyes, and heavy metals, even at low concentrations (Bhatnagar & Sillanpää, 2010).

Additionally, to reduce dependence on commercially activated carbon, agricultural residue, and industrial byproducts are often being used as low-cost natural adsorbents. However, adsorption has some critical limitations, as it is a non-destructive method that only transfers contaminants from liquid to solid phase, requiring safe disposal or regeneration of spent adsorbents. Activated carbon thermal regeneration requires a large amount of power and high operation costs, and the sequential regeneration stages usually result in a reduction in adsorption capacity due to pore clogging (Metcalf & Eddy/AECOM, 2014). The effectiveness of adsorption is also highly influenced by the wastewater characteristics, including the pH, ionic strength and the presence of other solutes, which may significantly affect contaminant uptake.

Electrochemical methods (e.g., electrocoagulation, electro-oxidation, and electro-Fenton reactions) utilize electrical current to destabilize pollutants, encourage coagulation, or generate reactive species that can mineralize organic contaminants. These methods are also particularly suitable for industrial wastewater and pollutants like dyes, phenols, and heavy metals (Martínez-Huitle & Ferro, 2006). This method has the major advantage of operational flexibility and precision, with the minimum addition of chemicals. Less sludge is generated from electrocoagulation than chemical coagulation, while electro-oxidation can mineralize organics into harmless end products such as CO_2 and water (Pera-Titus *et al.*, 2005). Moreover, the modular design of the electrochemical reactors allows the decentralized applications of treatment (Metcalf & Eddy/AECOM, 2014). This technique is however limited by the high operational costs from electricity consumption and electrode wear. Anode passivation minimizes long-term efficiency and chloride ions lead to the formation of toxic products such as chlorinated organics (Martínez-Huitle & Ferro, 2006). Furthermore, its applicability is restricted to certain types of wastewater depending on the wastewater composition.

Comparative study indicates that thermal treatment demonstrates superior results in eliminating pathogens but is hindered by high energy costs; adsorption provides simplicity and cost-effectiveness but merely transfers pollutants without degradation; and electrochemical degradation offers precision and modularity but remains limited by electricity demand and electrode maintenance. None of these techniques alone is sufficient to deal with modern wastewater, which is a mixture of new compounds like pharmaceuticals, microplastics and endocrine-disrupting compounds (Bhatnagar & Sillanpää, 2010). Recent research trends are on hybridization, for example, combining electrochemical oxidation with adsorption to regenerate used adsorbents, or using thermal hydrolysis before biological treatment to increase biodegradability (Yavuz, 2007). These integrated systems offer the strengths of individual methods while mitigating their limitations.

2.3 EVOLUTION TOWARD SMART TECHNOLOGIES

The evolution of smart technologies has accelerated due to the limitations of conventional wastewater management. These latest technologies leverage automation, real-time monitoring, and advanced analytics to improve system performance. Further integration of the Internet of Things (IoT), artificial intelligence (AI), and information and communication technologies (ICT) in smart wastewater systems enables continuous monitoring, data acquisition, predictive analysis, and efficient management of collection and treatment networks (Yavuz, 2007). For example, IoT-enabled sensors installed inside sewer networks provide real-time data on flow, pH, turbidity, and pollutant levels, enabling timely detection of anomalies such as blockages, illicit discharges, or combined sewer overflows (Mukherjee & Singh, 2020). ICT platforms empower operators and users with centralized information dashboards and useful insights. AI and machine learning algorithms enhance treatment plant performance via timely indications of pollutant loads, and support proactive maintenance scheduling (Wang *et al.*, 2021).

The adoption of digital twins–virtual replicas of wastewater networks–has further expanded smart wastewater management. By simulating hydraulic and biological processes in real time, digital twins support scenario testing and enhance system resilience against climate-induced variability (Schütze *et al.*, 2019). Similarly, blockchain-based platforms are being explored to improve data security and transparency in wastewater operations (Shi, 2020). These advances mark a shift from reactive treatment to proactive, data-driven wastewater management, enabled by real-time monitoring and advanced analytics. In the face of growing pressure from urbanization, climate change, and complex pollutants, smart wastewater technologies integrating IoT, AI, and digital platforms are essential for achieving resilient and sustainable water systems.

2.4 ROLE OF SMART TECHNOLOGIES IN WASTEWATER COLLECTION

Wastewater collection networks (WCNs) are essential urban infrastructure but have traditionally operated passively with reactive maintenance and limited real-time insight. Growing pressures from urbanization, climate change, and emerging pollutants have exposed the limitations of conventional systems. To overcome these challenges, smart technologies such as smart sensors, IoT, and ICT are being integrated to enable enhanced monitoring, control, and data-driven decision-making. Continuous, real-time sensing of key parameters provides higher spatial and temporal resolution than traditional sampling, forming the backbone of intelligent wastewater management (Yavuz, 2007). For example, ultrasonic level sensors are used to detect sewer overflow risk, while electrochemical sensors measure heavy metals and nutrient concentrations directly in sewer pipes. Optical and biosensors have been developed to detect pathogens and emerging contaminants such as pharmaceuticals at trace levels (Herrera-Domínguez *et al.*, 2023). The miniaturization and low-cost manufacturing of sensors have further accelerated their deployment across decentralized sewer systems. Despite these advances, challenges remain. Sensors deployed in harsh sewer environments suffer from biofouling, calibration drift, and energy constraints (Wang *et al.*, 2021). Maintenance and replacement costs are often high, and data reliability may degrade without robust validation protocols. Thus, while smart sensors extend system visibility, their long-term sustainability depends on advances in materials, energy harvesting, and self-calibration mechanisms.

2.5 INFORMATION AND COMMUNICATION TECHNOLOGIES

ICTs form the backbone of smart wastewater management, leveraging technologies such as cloud computing, data analytics, and geographic information systems (GIS). They process data from IoT sensors to provide actionable insights, enabling utilities to model wastewater flow and monitor systems more effectively. ICTs support decision-making during storm events by integrating hydraulic models and weather data to prevent flooding and sewer overflows. While fostering advanced data analytics for detecting illicit discharges, the implementation of ICT faces challenges, including the need for infrastructure investment, data governance, and issues like privacy and interoperability. Despite these challenges, ICT systems facilitate a shift toward proactive management, improving environmental compliance, energy efficiency, and resilience to climate variability (Wang *et al.*, 2022; Yavuz, 2007).

Smart technologies are transforming wastewater collection by replacing traditional manual systems with advanced intelligent networks. These systems incorporate smart sensors for real-time monitoring of water quality and flow parameters, IoT platforms for seamless data exchange and predictive maintenance, and ICT tools that support data analytics and decision-making processes. While challenges such as cost, cybersecurity, interoperability, and maintenance persist, the incorporation of these technologies leads to more sustainable, resilient, and efficient wastewater collection systems.

2.6 REMOTE SENSING APPLICATIONS IN WASTEWATER MANAGEMENT

Wastewater management using new techniques such as remote sensing (RS) involves utilization of new techniques for capturing, processing, and transmitting data on solid waste in a prompt and efficient manner. RS combined with GIS supports effective waste disposal which is essential for public health, environmental protection, and sustainable growth (Fu *et al.*, 2010). Significant increases in wastewater generation are due to increasing population density, industrial activities, and commercial establishments (Deselaers *et al.*, 2010). Data collection and analysis are now performed using statistical methods (Čampulová *et al.*, 2018), machine learning techniques (Akouemo & Povinelli, 2017), GIS, and deep learning approaches (An *et al.*, 2020), offering comprehensive time-series monitoring and significantly enhanced management of wastewater systems (Figure 2.2).

RS facilitates early detection of pollution, infrastructure monitoring, and enhances treatment efficiency. This enables the monitoring of water quality parameters such as turbidity, chlorophyll-a, total suspended solids, and nutrient concentrations. Satellite-based platforms, including optical and microwave sensors, provide spatiotemporal data for early warning systems of contaminants and pollutants in water. Further utilization of multispectral image analysis with machine learning enhances pollutant detection and classification (Adjovu *et al.*, 2023; Gålfalk *et al.*, 2022). Various spaceborne are used for monitoring wastewater parameters, with many US-based satellite datasets being available

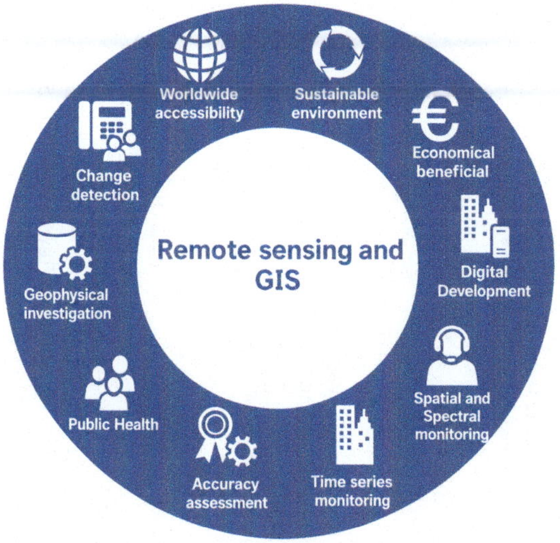

Figure 2.2 Features of RS and GIS for wastewater management.

globally free of cost. Different sensors offer different spatial (surface area), spectral (the sensor's ability to resolve features in the electromagnetic spectrum), radiometric, and temporal resolutions. Some widely used sensors for water quality monitoring are given in Table 2.1.

RS detects two optical parameters, that is, water color and turbidity, which impact the energy levels detected by the sensors. Water color is detected by a multispectral scanner camera, but its accuracy is challenging due to many variables governing energy level measure by sensor, which includes the time of year, sun elevation angle, aerosol and trace gas concentrations in the atmosphere, water-vapor content of the atmosphere, specular reflection of skylight from the water surface, and so on.

GIS-based spatial tools such as Multi Criteria Decision Analysis (MCDA) and analytic hierarchy process (AHP) focus on the identification, tracking, and monitoring of water pollution sites in water bodies and land-use changes affecting wastewater systems. GIS integrated in wastewater treatment plants can identify sewage leakage, flooding, and contaminant zones. It provides a spatial framework for integrating, visualizing, and analyzing data related to WWT plants via AHP methods. Further automation by integrating machine learning models can be done, which deals with the factorial relationship among components which influence demand on wastewater treatment infrastructure. This identifies and monitors sources of wastewater discharge and water bodies influencing wastewater systems.

Cusworth *et al.* (2020) utilized the Next Generation Airborne Visible/Infrared Imaging Spectrometer (AVIRIS-NG) for quantifying strong methane emissions from waste management by California Methane Survey who

Table 2.1 Widely used remote sensors in water quality monitoring.

Sensors	Launched	Features	References
Landsat satellite images	USGS and NASA for earth observation have been in existence since 1972.	Land use planning, natural resources management, public safety, climate research, natural disaster management, home security, and agriculture	Nxumalo *et al.* (2026)
Landsat 9 OLI/TIRS	27 September 2021	Latest series of earth-observing satellites Operational Land Imager 2 (OLI–2) and the Thermal Infrared Sensor 2 (TIRS–2) Nine spectral bands Wider synoptic global earth's surface coverage for the detection of natural and human-induced spatiotemporal scale changes	Pettorelli (2019)
Landsat 8 OLI/TIRS	11 February 2013	Nine spectral bands spatial resolution of 30 m	El-Zeiny and Effat (2017), Adjovu *et al.* (2023)
Landsat 7 Enhanced Thematic Mapper (ETM+)	15 April 1999	Landsat 7 images are delivered in 8 bits with 256 gray levels 30 m resolution all bands, apart from band 6 and band 8 with 60 and 15 m	Adjovu *et al.* (2023)
RapidEye sensor	29 August 2008	6.5 m spatial resolution 77 km swath width, 5.5-day revisit time, 11:00 AM equator crossing, 12-bit dynamic range, and 5 million km^2/day image capture capacity with 7-year spacecraft lifetime. Retirement on 31 March, 2020	Tewes *et al.* (2015)

Advanced Spaceborne Thermal Emission and Reflection Radiation (ASTER)	December 1999	Spatial resolutions of 15, 30, and 90 m. Comprises Visible and Near-infrared (VNIR), SWIR, and Thermal Infrared (TIR) subsystem	Abrams *et al.* (1999)
Moderate Resolution Imaging Spectroradiometer (MODIS)	December 1999 (Terra) and May 2002 (Aqua)	36 spectral bands with varying wavelengths and spatial resolution	Singh *et al.* (2021)
Sentinel-2 MSI	23 June 2015, and 7 March 2017 (Sentinel 2A and 2B)	10 days revisit time with one satellite and 5 days revisit time with the two satellites. Study of water quality, impervious surface mapping, monitoring of land ecosystems and land-use land cover (LULC), forest management, agriculture, and disaster mapping	Pahlevan *et al.* (2017)

flew over 270 landfills and 166 organic waste facilities repeatedly between 2016 and 2018. Alterations in emissions resulted in identifying changes in landfill infrastructure (design, construction, and operations). Using imaging spectrometers to help quantify the total methane emission for any given landfill is challenging, as operations, meteorology, topography, and infrastructure change constantly.

Pandey *et al.* (2022) suggested utilization of GIS and water quality indexes to determine the quality of groundwater through physico-chemical parameters for drinking purposes. The contaminants are elevated in concentration due to the exploitation of groundwater and anthropogenic activities.

2.7 IOT-BASED WASTEWATER COLLECTION SYSTEMS

The IoT connects distributed sensors, meters, and actuators across wastewater networks, allowing seamless data acquisition, transmission, and analysis. IoT platforms integrate heterogeneous devices into a common architecture, providing centralized dashboards for utilities and municipalities. For instance, wireless sensor networks (WSNs) transmit real-time data on sewer flow and water quality to cloud-based platforms, where machine learning algorithms can identify anomalies or predict failure risks (Mukherjee & Singh, 2020).

IoT also enables predictive and prescriptive maintenance, where smart pumps and valves equipped with IoT controllers can automatically adjust operation based on flow dynamics or pollutant loads, reducing energy use and preventing overflow events. Case studies in European cities have shown that IoT-enabled sewer monitoring reduced combined sewer overflows (CSOs) by more than 30% through proactive interventions (Garcia *et al.*, 2020).

However, IoT adoption faces limitations related to cybersecurity, interoperability, and data overload. Ensuring secure communication protocols is critical, as wastewater infrastructure is increasingly recognized as a target for cyberattacks. Furthermore, integration across multiple vendors' devices requires adherence to open communication standards, which remain inconsistent across regions (Shi, 2020).

IoT-based wastewater collection systems require centralized control units and WSNs to operate the system. They utilize data obtained from sensors and track the color of wastewater fluid, atmospheric conditions, and temperature of the treatment plant. This IoT-based approach detects and avoids sewage mixing and the overflow of chemicals in wastewater. An IoT-based smart wastewater system can fulfil the demand for freshwater in a locality and help increase freshwater availability. IoT-based sensors installed across water resources can play an important role in saving aquatic life (Amin *et al.*, 2022). These sensors act as a bridge between surveillance systems and water bodies, generating datasets such as water quality, temperature and pressure changes, and leakage detection, which helps in monitoring water quality, pressure, and temperature. This block chain technology efficiently monitors connections enabling timely rectification of breaches and leakage in the system.

Landa-Cansigno *et al.* (2020) analyzed different approaches for efficiency evaluation of water reuse and suggested an integrated framework for urban water metabolism (UWM) and the water–energy–pollution nexus (WEPN) illustrating the interaction among the nexus (e.g., water, energy, and pollution) components.

2.7.1 IoT-WMS (water metabolism framework)

IoT-WMS is an established smart wastewater collection system for smart cities focused on cloud protection and leveraging blockchain technologies. IoT-WMS uses a trading scheme focused on the recovery of wastewater using blockchain incentives. This approach is available to all household and industries through the utilization of sensors and actuators. This provides a high wastewater recycling rate of 96.3%, an efficiency ratio of 88.7%, a low moisture content ratio of 32.4%, an increased wastewater reuse of 90.8%, and a prediction ratio of 92.5%. However, this approach cannot be used in rural areas and faces inoperability challenges between the different systems deployed by a number of metropolitan and/or rural areas (Alzahrani *et al.*, 2023).

2.7.2 IoT-based radar sensor network

WCNs are often ignored and rarely observed but they are a major component for managing waste from homes to sewage treatment. Drenoyanis *et al.* (2019) developed a prototype device utilizing the LoRaWAN protocol, a sewer pumpstation universal drawdown device (SPUDD) and a LoRaWAN IoT-network enabled device. Additionally, signal strength measurements were built to perform (radio channel ranging survey). In addition to providing a useful engineering solution for efficiently monitoring sewer pumpstation pump performance, these devices also serve as a portable first solution for long-term gravity sewer system monitoring by indirectly estimating corresponding pump flows within sewer pumpstation collection wells. LoRaWAN has been proved to be low power, effective and wide area network in urban areas.

2.7.3 Supervisory control and data acquisition (SCADA) system

The efficient functioning of a sewer system is crucial as it reduces the possibility of malfunction and overflow. The integration of IoT with a SCADA system enhances monitoring and control by combining Deep Belief Networks (DBNs). An integrated IoT-SCADA system can effectively oversee multiple phases of wastewater collection and treatment processes. By integrating DBNs, the framework gains advanced predictive analytics that support preventive maintenance strategies and data-driven decision-making, helping avoid system failures while maximizing efficient resource use. This integrated approach maintains chemical discharge levels within compliance standards and provides ongoing assessment of treatment facility performance. Additionally, Complex Event Processing (CEP) technology enables the system to handle and analyze the vast volumes of real-time sensor data, facilitating immediate insights and dynamic operational adjustments.

IoT-based technologies enable the detection of humidity, temperature, and gases mixture levels in sewage water which track real-time dynamic changes in the above factors. This gives timely updates to the user remotely via mobile devices, in case any blockage is encountered or contaminant level is above threshold level (Pendharkar *et al.*, 2020).

2.8 AUTOMATED WASTEWATER COLLECTION AND TRANSPORT

Automation is emerging as multifaceted technological advancement incorporating AI, robotics, sensors, and cloud computing. Deep learning approaches have transformed wastewater collection systems into automated systems. Most popular CNN-based algorithms are utilized for pipe defect classification (Goodfellow *et al.*, 2016; Gu *et al.*, 2018). This algorithm uses both supervised and unsupervised learning approaches to minimize complex feature extraction and training processes, providing higher accuracy and less processing speed.

An approach common worldwide for the detection of leaks in sewer pipes is by utilizing closed-circuit television (CCTV); the images are processed manually which is time-consuming and labor-intensive. An automated approach was developed for detecting sewer pipes based on a deep learning approach and model development. Cheng and Wang (2018) developed a detection model by training CCTV inspection based on 3000 images of sewer pipes. This approach has higher precision and accuracy, so it has the capability of reducing inspection time and resources, as well as enhancing the accuracy of the sewage leakage detection model.

Virtual optimization and wastewater treatment processes can be achieved by re-fitting bio-kinetics models (soft-sensors) using data-driven models. These models enable the visualization of continuous future process trajectories, allowing the prediction of discrete data and facilitating precise monitoring and control of system activity (Li *et al.*, 2019). This transformation is further accelerated by the digital revolution encompassing big data, data analytics, AI, digital tools such as sensors, digital twins, data mining, data pipelines, the IoT, water surveillance systems, and nanotechnology in the development of water treatment functionalities.

The growing availability of affordable sensors, high-resolution RS, and advanced communication technologies (such as wireless networks and optical fiber) is fuelling the generation of big data in the water sector and strengthening decision-making processes. The integration of big data and AI allows for the analysis, structuring, and extraction of actionable insights from large and diverse datasets, opening up data-driven opportunities to optimize the operation and control of wastewater treatment systems (Blumensaat *et al.*, 2019; Dayanidhi *et al.*, 2020).

2.9 CHALLENGES IN SMART WASTEWATER INFRASTRUCTURE

The implementation of smart wastewater infrastructure is significantly influenced by organizational factors, not just technical challenges. The involvement of the local population is crucial for the success of these initiatives. Although advanced technologies such as IoT and AI present transformative opportunities for wastewater management, their adoption is impeded by various challenges, including sensor accuracy, energy efficiency, communication reliability, and cybersecurity concerns. Financial limitations, characterized by substantial initial investments and continual operational expenses, particularly affect developing regions. Additionally, outdated infrastructure complicates

Table 2.2 Challenges faced in smart wastewater techniques implementation.

Challenges	Description	References
Technology Integration	Sensor accuracy, IoT/AI adoption, interoperability	Alshami *et al.* (2024), Cairone *et al.* (2024)
Financial Barriers	High costs, limited funding, investment gaps	Dominguez and Gujer (2006), Liang and Yue (2021)
Cybersecurity	Vulnerability to attacks, lack of expertise	Pandey *et al.* (2022)
Governance	Inefficient management, lack of collaboration	Liang and Yue (2021)
Public Acceptance	Resistance to water reuse, awareness gaps	Villarín and Merel (2020)

upgrades, raising issues related to equipment durability and frequent failures. Governance and social factors are also critical, necessitating interdisciplinary collaboration, public acceptance of water reuse, and addressing the imbalance in technical expertise and resources between developed and developing areas. Environmental pressures, including the impacts of climate change such as flooding and variable treatment performance, add further layers of complexity to maintaining reliable operations as depicted in Table 2.2. Overcoming these challenges requires comprehensive strategies that integrate technological innovation, enhance workforce capacity, establish supportive policy frameworks, and promote increased stakeholder engagement.

2.10 CONCLUSION

The transition toward a digitized wastewater management ecosystem represents a significant shift in how global water scarcity is tackled. By leveraging IoT sensors, RS, and ICT, wastewater is transformed into a valuable resource instead of a mere byproduct, establishing a real-time monitoring 'nervous system' for urban water infrastructure. The integration of AI enhances data utilization for predictive analytics, while blockchain ensures transparency in water reuse tracking. This modern approach aligns with the UN Sustainable Development Goals by improving water quality and promoting circular economy practices, essential for countering climate change. However, the global implementation of these technologies faces substantial obstacles, notably high capital costs which contribute to a 'digital divide' in developing regions. Additionally, the absence of standardized frameworks leads to fragmented and unmanageable systems, while cybersecurity risks increase as critical infrastructure becomes digitized. Furthermore, practical scalability issues arise from sensor energy demands, limited wireless coverage, and weak IoT infrastructure in resource-limited settings.

To achieve global water security, the next steps must focus on developing affordable and well-integrated solutions customized for various contexts. This technical advancement should coincide with supportive policy frameworks and collaborative partnerships between technology providers and public utilities. Emphasizing capacity building and equitable access can convert smart wastewater management from an expensive luxury into an essential sustainability tool. Addressing these challenges is crucial for leveraging modern technology to alleviate the global water crisis and securing a resilient water future for future generations.

REFERENCES

Abrams M., Hook S. and Ramachandran B. (1999). ASTER user handbook, version 2. Jet Propulsion Laboratory, Pasadena, CA and EROS Data Center, Sioux Falls, SD.

Adjovu G. E., Stephen H., James D. and Ahmad S. (2023). Overview of the application of remote sensing in effective monitoring of water quality parameters. *Remote Sensing*, **15**, 1938, https://doi.org/10.3390/rs15071938

Akouemo H. N. and Povinelli R. J. (2017). Data improving in time series using ARX and ANN models. *IEEE Transactions on Power Systems*, **32**, 3352–3359, https://doi.org/10.1109/TPWRS.2017.2656939

Alshami A., Ali E., Elsayed M., Eltoukhy A. E. and Zayed T. (2024). IoT innovations in sustainable water and wastewater management and water quality monitoring: a comprehensive review of advancements, implications, and future directions. *IEEE Access*, **12**, 58427–58453, https://doi.org/10.1109/ACCESS.2024.3392573

Alzahrani A. I., Chauhdary S. H. and Alshdadi A. A. (2023). Internet of Things (IoT)-based wastewater management in smart cities. *Electronics*, **12**, 2590, https://doi.org/10.3390/electronics12122590

Amin F., Abbasi R., Mateen A., Ali Abid M. and Khan S. (2022). A step toward next-generation advancements in the Internet of Things technologies. *Sensors*, **22**, 8072, https://doi.org/10.3390/s22208072

An Q., Tao Z., Xu X., El Mansori M. and Chen M. (2020). A data-driven model for milling tool remaining useful life prediction with convolutional and stacked LSTM network. *Measurement*, **154**, 107461, https://doi.org/10.1016/j.measurement.2019.107461

Angelakis A. N., Asano T., Bahri A., Jimenez, B. E. and Tchobanoglous, G. (2018). Water reuse: from ancient to modern times and the future. *Frontiers in Environmental Science*, **6**, 26, https://doi.org/10.3389/fenvs.2018.00026

Bhatnagar A. and Sillanpää M. (2010). Utilization of agro-industrial and municipal waste materials as potential adsorbents for water treatment–A review. *Chemical Engineering Journal*, **157**(2–3), 277–296, https://doi.org/10.1016/j.cej.2010.01.007

Blumensaat F., Leitão J. P., Ort C., Rieckermann J. R., Scheidegger A., Vanrolleghem P. A. and Villez K. (2019). How urban storm-and wastewater management prepares for emerging opportunities and threats: digital transformation, ubiquitous sensing, new data sources, and beyond–A horizon scan. *Environmental Science & Technology*, **53**, 8488–8498, https://doi.org/10.1021/acs.est.8b06481

Boretti A. and Rosa L. (2019). Reassessing the projections of the world water development report. *Nature Sustainability*, **2**(9), 873–875, https://doi.org/10.1038/s41545-019-0039-9

Cairone S., Hasan S. W., Choo K.-H., Lekkas D. F., Fortunato L., Zorpas A. A., Korshin G., Zarra T., Belgiorno V. and Naddeo V. (2024). Revolutionizing wastewater treatment toward circular economy and carbon neutrality goals: pioneering sustainable and efficient solutions for automation and advanced process control with smart and cutting-edge technologies. *Journal of Water Process Engineering*, **63**, 105486, https://doi.org/10.1016/j.jwpe.2024.105486

Čampulová M., Michálek J., Mikuška P. and Bokal D. (2018). Nonparametric algorithm for identification of outliers in environmental data. *Journal of Chemometrics*, **32**, e2997, https://doi.org/10.1002/cem.2997
Cheng J. C. and Wang M. (2018). Automated detection of sewer pipe defects in closed-circuit television images using deep learning techniques. *Automation in Construction*, **95**, 155–171, https://doi.org/10.1016/j.autcon.2018.08.006
Cusworth D. H., Duren R. M., Thorpe A. K., Tseng E., Thompson D., Guha A., Newman S., Foster K. T. and Miller C. E. (2020). Using remote sensing to detect, validate, and quantify methane emissions from California solid waste operations. *Environmental Research Letters*, **15**, 054012, https://doi.org/10.1088/1748-9326/ab7b99
Dayanidhi K., Vadivel P., Jothi S. and Eusuff N. S. (2020). Facile synthesis of Silver@ Eggshell nanocomposite: A heterogeneous catalyst for the removal of heavy metal ions, toxic dyes and microbial contaminants from water. *Journal of Environmental Management*, **271**, 110962, https://doi.org/10.1016/j.jenvman.2020.110962
Deselaers T., Heigold G. and Ney H. (2010). Object classification by fusing SVMs and Gaussian mixtures. *Pattern Recognition*, **43**, 2476–2484, https://doi.org/10.1016/j.patcog.2010.02.002
Dominguez D. and Gujer W. (2006). Evolution of a wastewater treatment plant challenges traditional design concepts. *Water Research*, **40**, 1389–1396, https://doi.org/10.1016/j.watres.2006.01.034
Drenoyanis A., Raad R., Wady I. and Krogh C. (2019). Implementation of an IoT based radar sensor network for wastewater management. *Sensors*, **19**, 254, https://doi.org/10.3390/s19020254
El-Zeiny A. M. and Effat H. A. (2017). Environmental monitoring of spatiotemporal change in land use/land cover and its impact on land surface temperature in El-Fayoum governorate, Egypt. *Remote Sensing Applications: Society and Environment*, **8**, 266–277, https://doi.org/10.1016/j.rsase.2017.10.003
Fu Q., Wang Z. and Jiang Q. (2010). Delineating soil nutrient management zones based on fuzzy clustering optimized by PSO. *Mathematical and Computer Modelling*, **51**, 1299–1305, https://doi.org/10.1016/j.mcm.2009.10.034
Gålfalk M., Påledal S. N., Sehlén R. and Bastviken D. (2022). Ground-based remote sensing of CH_4 and N_2O fluxes from a wastewater treatment plant and nearby biogas production with discoveries of unexpected sources. *Environmental Research*, **204**, 111978, https://doi.org/10.1016/j.envres.2021.111978
García L., Parra L., Jimenez J. M., Lloret J. and Lorenz P. (2020). IoT-based smart irrigation systems: An overview on the recent trends on sensors and IoT systems for irrigation in precision agriculture. *Sensors*, **20**, 1042, https://doi.org/10.3390/s20041042
Goodfellow I., Bengio Y. and Courville A. (2016). Deep Learning. MIT Press, Cambridge, MA.
Gu J., Wang Z., Kuen J., Ma L., Shahroudy A., Shuai B., Liu T., Wang X., Wang G., Cai J. and Chen T. (2018). Recent advances in convolutional neural networks. *Pattern Recognition*, **77**, 354–377. https://doi.org/10.1016/j.patcog.2017.10.013
Herrera-Domínguez M. Morales-Luna G., Mahlknecht J., Cheng Q., Aguilar-Hernández I. and Ornelas-Soto N. (2023). Optical biosensors and their applications for the detection of water pollutants. *Biosensors*, **13**, 370, https://doi.org/10.3390/bios13030370
Landa-Cansigno O., Behzadian K., Davila-Cano D. I. and Campos L. C. (2020). Performance assessment of water reuse strategies using integrated framework of urban water metabolism and water-energy-pollution nexus. *Environmental Science and Pollution Research*, **27**, 4582–4597, https://doi.org/10.1007/s11356-019-05465-8

Li X., Mei Q., Chen L., Zhang H., Dong B., Dai X., He C. and Zhou J. (2019). Enhancement in adsorption potential of microplastics in sewage sludge for metal pollutants after the wastewater treatment process. *Water Research*, **157**, 228–237, https://doi.org/10.1016/j.watres.2019.03.069

Liang X. and Yue X. (2021). Challenges facing the management of wastewater treatment systems in Chinese rural areas. *Water Science and Technology*, **84**, 1518–1526. https://doi.org/10.2166/wst.2021.332

Martinez-Huitle C. A. and Ferro S. (2006). Electrochemical oxidation of organic pollutants for wastewater treatment: direct and indirect processes. *Chemical Society Reviews*, **35**, 1324–1340. https://doi.org/10.1039/B517632H.

Metcalf & Eddy/AECOM (2014). Wastewater Engineering: Treatment and resource recovery. McGraw-Hill Education, New York.

Mohan D. and Pittman Jr C. U. (2006). Activated carbons and low cost adsorbents for remediation of tri- and hexavalent chromium from water. *Journal of Hazardous Materials*, **137**, 762–811. https://doi.org/10.1016/j.jhazmat.2006.06.060

Mukherjee I. and Singh U. K. (2020). Delineation of groundwater potential zones in a drought-prone semi-arid region of east India using GIS and analytical hierarchical process techniques. *Catena*, **194**, 104681. https://doi.org/10.1016/j.catena.2020.104681

Nxumalo N., Nzimande N. P. and Xulu S. (2026). Remote Sensing Applications for Land-Use and Land-Cover Change Research in South African Landscapes: A Review. *Earth*, **7**, 54, https://doi.org/10.3390/earth7020054

Pandey S., Twala B., Singh R., Gehlot A., Singh A., Montero E. C. and Priyadarshi N. (2022). Wastewater treatment with technical intervention inclination towards smart cities. *Sustainability*, **14**, 11563, https://doi.org/10.3390/su141811563

Pahlevan N., Sarkar S., Franz B. A., Balasubramanian S. V. and He J. (2017). Sentinel-2 MultiSpectral Instrument (MSI) data processing for aquatic science applications: Demonstrations and validations. *Remote Sensing of Environment*, **201**, 47–56, https://doi.org/10.1016/j.rse.2017.08.033

Pera-Titus M., Llorens J., Cunill F., Mallada R. and Santamaría J. (2005). Preparation of zeolite NaA membranes on the inner side of tubular supports by means of a controlled seeding technique. *Catalysis Today*, **104**, 281–287. https://doi.org/10.1016/j.cattod.2005.03.042

Pendharkar A., Chillapalli J., Dhakate K., Gogoi S. and Jadhav Y. (2020). IoT based sewage monitoring system. *Proceedings of the International Conference on Recent Advances in Computational Techniques (IC-RACT)*, http://dx.doi.org/10.2139/ssrn.3697395

Pettorelli N. (2019). Satellite remote sensing and the management of natural resources, Oxford University Press, https://doi.org/10.1093/oso/9780198717263.001.0001

Schütze M., Campisano A., Colas H., Schilling W. and Vanrolleghem P. A. (2019). Real time control of urban wastewater systems: Where do we stand today? *Journal of Hydrology*, **299**(3-4), 335–348. https://doi.org/10.1016/j.jhydrol.2004.08.010

Shi T., Han G., Ma X., Zhang M., Pei Z., Xu H., Qiu R., Zhang H. and Gong W. (2020). An inversion method for estimating strong point carbon dioxide emissions using a differential absorption Lidar. *Journal of Cleaner Production*, **271**, 122434, https://doi.org/10.1016/j.jclepro.2020.122434

Singh R. K., Sinha V. S. P., Joshi, P. K. and Kumar, M. (2021). A multinomial logistic model-based land use and land cover classification for the South Asian Association for Regional Cooperation nations using Moderate Resolution Imaging Spectroradiometer product. *Environment, Development and Sustainability*, **23**, 6106–6127, https://doi.org/10.1007/s10668-020-00864-1

Tewes A., Thonfeld F., Schmidt M., Oomen R. J., Zhu X., Dubovyk O., Menz G. & Schellberg J. (2015). Using RapidEye and MODIS data fusion to monitor vegetation dynamics in semi-arid rangelands in South Africa. *Remote Sensing*, **7**, 6510–6534, https://doi.org/10.3390/rs70606510

UN-Water (2021). *Progress on change in water-use efficiency: Global status and acceleration needs for SDG indicator* **6**.4.1, 2021. UN-Water Publications.

UNESCO World Water Assessment Programme (2019). The United Nations World Water Development Report 2019, Leaving No One Behind, p. 186. ISBN 978–92–3-100309–7, UNESCO, Paris, France.

Villarín M. C. and Merel S. (2020). Paradigm shifts and current challenges in wastewater management. *Journal of Hazardous Materials*, **390**, 122139. https://doi.org/10.1016/j.jhazmat.2020.122139

Wang M., Orr A. A., Jakubowski J. M., Bird K. E., Casey C. M., Hearon S. E., Tamamis P. and Phillips T. D. (2021). Enhanced adsorption of per- and polyfluoroalkyl substances (PFAS) by edible, nutrient-amended montmorillonite clays. *Water Research*, **188**, 116534, https://doi.org/10.1016/j.watres.2020.116534

Wang G., Jia Q. S., Zhou M., Bi J., Qiao J. and Abusorrah A. (2022). Artificial neural networks for water quality soft-sensing in wastewater treatment: a review. *Artificial Intelligence Review*, **55**, 565–587, https://doi.org/10.1007/s10462-021-10038-8

Yavuz Y. (2007). EC and EF processes for the treatment of alcohol distillery wastewater. *Separation and Purification Technology*, **53**(1), 135–140. https://doi.org/10.1016/j.seppur.2006.08.022

doi: 10.2166/9781789065404_0037

Chapter 3

Advanced and intelligent wastewater treatment technologies

Aftab Ahmad Khan[1], *Lovepreet Singh*[2]*, *Niketa Bhati*[3] and *Harshita Jain*[3]

[1]Department of Civil and Environmental Engineering, College of Engineering, King Faisal University, Al-Ahsa 31982, Saudi Arabia. E-mail: aabkhan@kfu.edu.sa
[2]Department of Chemical Engineering, Thapar Institute of Engineering and Technology, Patiala 147004, India. E-mail: lovepreet.singh@thapar.edu
[3]Department of Climate Change and Environmental Sustainability, Amity Institute of Environment and Sustainability, Amity University Uttar Pradesh (AUUP), Noida, Uttar Pradesh 201301, India. E-mail: hjain@amity.edu
**Corresponding author: lovepreet.singh@thapar.edu*

ABSTRACT

Recent advances in wastewater treatment technologies have encouraged a shift away from traditional process design toward systems more adaptable, more efficient and responsive to changing influent characteristics and operating requirements. The evolution of intelligent treatment strategies has been accelerated by regulatory pressures, higher treatment standards, and the need to decrease energy and chemical consumption. This chapter explores the evolution of modern wastewater treatment technologies, including the use of artificial intelligence, automation, and process optimization. In particular, it addresses new discoveries in biological, chemical and hybrid treatment, as well as the ways in which intelligent control approaches improve treatment performance, stability, and resource efficiency. Machine learning is presented, and models driven by data as tools for real-time monitoring, fault finding, and process control. The technology is shown to extend existing systems through informed decision making and adaptive practice rather than replacing existing treatment methods. Practical considerations such as system scalability, operational reliability, and implementation problems in both centralized and decentralized environments are also addressed.

Keywords: Advanced wastewater treatment, intelligent systems, artificial intelligence, process optimization, biological treatment, hybrid treatment technologies

3.1 INTRODUCTION

Traditional wastewater treatment systems have been a major source of public health and water quality protection for decades (Obaideen *et al.*, 2022). Primary sedimentation, activated sludge treatment, and secondary clarification processes were utilized to address relatively stable influent compositions and clearly defined regulatory goals. While these systems continue to be the foundation of municipal and industrial wastewater management, their design philosophy is still largely based on fixed operating conditions and linear process control. As the characteristics of wastewater get more complex and the pressures from the outside become more stringent, the limitations of existing treatment systems become increasingly clear (Abdelrahman *et al.*, 2023). Many existing plants operate at least as close as possible to their intended capacity, leaving little room for adaptation to fluctuation in flow rate, pollutant load or influent composition (Kolosovska & Bauer, 2022). Industrial discharges, new contaminants, and seasonal variability further complicate systems that were not constructed to dynamically adapt. These limits bring attention to the need to reevaluate the efficacy of traditional treatment practices in terms of the modern wastewater management needs (Gad *et al.*, 2022) (Table 3.1).

As highlighted by Yeasmin *et al.* (2023), another significant limitation of conventional wastewater treatment is that it requires steady-state operation.

Table 3.1 Limitations of conventional wastewater treatment systems under modern conditions.

Aspect	Conventional Treatment Characteristics	Key Limitations	References
Process control	Fixed setpoints and manual adjustments	Limited responsiveness to influent variability	Parsa *et al.* (2024)
Influent variability handling	Designed for average loading conditions	Performance deterioration during shock loads	Pophali *et al.* (2003)
Energy use	Conservative aeration and pumping strategies	High and often inefficient energy consumption	Gu *et al.* (2023)
Pollutant removal scope	Focus on bulk organics and solids	Limited removal of nutrients and micropollutants	Worku *et al.* (2025)
Monitoring	Periodic sampling and basic online sensors	Delayed detection of process disturbances	Moretti *et al.* (2024)
Sustainability alignment	Compliance-driven operation	Limited integration of energy recovery and circular approaches	Yusuf *et al.* (2026)

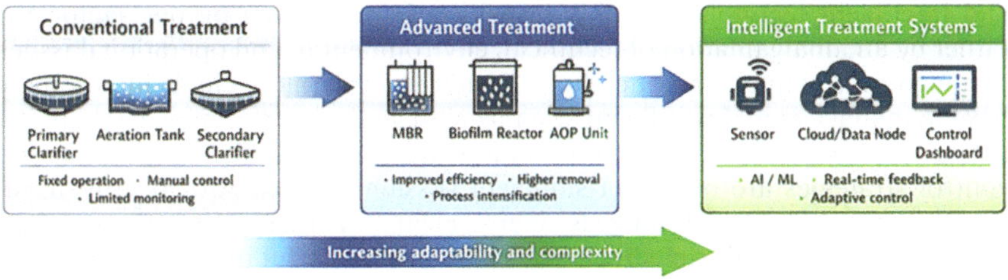

Figure 3.1 Evolution of wastewater treatment technologies toward intelligent systems.

Bio-based processes, such as activated sludge systems, are sensitive to organic loading, temperature, and hydraulic retention time. Influence quality change can be sudden and destabilize microbial activity, resulting in reduced treatment efficiency, sludge bulking, or process instability (Das & Mishra, 2025). In many cases, operators employ manual adjustment and empirical experience to make corrective actions rather than predictive tools that can retard correction and increase the chance of compliance failure (Shah, 2024; Shah & Rodriguez-Couto, 2019). Inefficiencies also exist in energy and chemical consumption. Aeration systems are a major source of energy used by wastewater treatment plants and these systems are often operated conservatively in order to avoid underperformance, which in turn causes excessive energy consumption. Similarly, chemical dosing for nutrient removal or pH control is frequently based on fixed setpoints rather than real-time process conditions (Jain & Shah, 2025). These practices contribute to higher operating costs and limit opportunities for resource recovery and optimization (Obaideen *et al.*, 2022; Raabe, 2023) (Figure 3.1).

As noted by Pratap *et al.* (2023), water quality, water reuse, and long-term sustainability have been the dominant factors in determining wastewater discharges. Increasing discharge limits for nutrients, pathogens, and trace organic pollutants have increased and placed extra strain on treatment systems that were originally designed to meet lower standards. The conventional treatment approaches are not always able to adequately address low levels of nutrients or emerging contaminants such as pharmaceuticals, personal care products, and microplastics (Satyam & Patra, 2024). Similarly, environmental goals shift beyond effluent quality to energy efficiency, greenhouse gases, and overall life cycle impacts of treatment infrastructure. The outdated, unreliable systems lack control and monitoring capabilities, and have little capacity to balance sustainability goals and compliance. These regulatory and environmental factors show the need for treatment technologies that can dynamically respond to changes in needs while remaining secure and efficient (Singh *et al.*, 2025).

3.2 DRIVERS FOR ADVANCED AND INTELLIGENT TREATMENT TECHNOLOGIES

As discussed by Zamfir *et al.* (2025), the increasing limitations of standard wastewater treatment systems led to the development of more sophisticated and intelligent technologies. This transition is not driven by one factor but rather by an amalgamation of technical, environmental, and operational issues that modern treatment facilities face. Wastewater treatment plants increasingly have to manage complex influent streams with greater performance, reliability, and sustainability. Under such conditions, static process designs and rule-based control strategies are often not sufficient (Essamlali *et al.*, 2024). Advanced, intelligent treatment technologies offer a means of interpreting process data more effectively, are flexible to changes and may be useful in making informed operating decisions. Their use, of course, is reflective of wider recognition that wastewater treatment systems should move away from 'tall infrastructure' into 'software-based, fully-responsive platforms that can match the performance of treatment with resource efficiency' (Palermo *et al.*, 2022).

3.2.1 Rising treatment complexity and variable influent loads

As per Nishmitha *et al.* (2025), the environmental properties of the wastewater have become more heterogeneous as a consequence of urban development, industrial diversification, and the introduction of new chemical compounds into domestic and industrial streams. Influent wastewater contains increasingly complex mixtures of organic, nutrients, heavy metals, and trace contaminants, many of which accumulate in high concentrations over relatively short time periods. Also, climatic variables such as high rainfall and prolonged dry periods also cause variability in hydraulic and pollutant loading (Fernandes *et al.*, 2024). Conventional treatment systems are typically designed for most conditions and are not flexible enough to deal effectively with rapid or unexpected changes. So, for example, influent conditions that deviate from the design assumptions may cause process upsets, lower removal efficiencies and operational instability (Kabir & Anick, 2024). High-tech treatment technologies coupled with intelligent monitoring and control techniques enable treatment systems to identify changes in influent quality and control operational parameters accordingly. This type of adaptive capacity is especially important in those facilities that operate within mixed municipal and industrial catchments, where influent variability is both frequently present and is difficult to predict (Yang *et al.*, 2025).

3.2.2 Energy efficiency and operational sustainability

As emphasized by Ghimire *et al.* (2021), for wastewater treatment, energy consumption is one of the most significant operational issues, with a large proportion of energy used for aeration, pumping, and sludge handling. New energy technologies and increasing emphasis on carbon reduction have driven wastewater utilities to seek ways to save money while maintaining good treatment efficiency. Conventional operating strategies often rely on conservative safety margins, which can lead to persistent over-aeration, excessive chemical use, and avoidable energy losses. Advanced and intelligent treatment technologies support more precise control of energy-intensive processes by linking real-time process data with predictive models and optimization algorithms. By aligning operational inputs with actual treatment demands, these systems can reduce energy consumption while maintaining regulatory goals (Nishmitha *et al.*, 2025). In addition, intelligent frameworks facilitate the integration of resource recovery strategies, such as energy generation from biogas or nutrient reuse, contributing to a more sustainable and circular approach to wastewater management (Gupta & Khatiwada, 2024).

3.3 ADVANCED BIOLOGICAL TREATMENT PROCESSES

As noted by Silva (2023), the biological treatment continues to be a staple of wastewater treatment because of its effectiveness, versatility, and ability to recycle a wide variety of organic and nutritional pollutants (Table 3.2). Reducing the complexity of the wastewater streams and increasing the standards of discharge, biological treatment technologies have been re-engineered. Design, microbial process control, and integration of modern biological systems have provided a wider performance envelope that allows conventional biological systems to function more efficiently and reliably in increasingly tight conditions. Sravan *et al.* (2024), found that, the advanced biological treatment processes allow better retention of biomass, improve microbial activity, and make processes more stable. These improvements lead to higher loading rates, improved effluent quality, and smaller operating footprints. In combination with intelligent monitoring and control, advanced biological systems can provide scalable solutions for both centralized and decentralized wastewater treatment applications (Figure 3.2).

Table 3.2 Performance characteristics of advanced biological treatment processes.

Process Type	Key Features	Typical Removal Efficiency	Energy Demand	Major Applications	References
Enhanced activated sludge	Optimized aeration, nutrient removal configurations	COD: 85–95%; TN: 60–80%	Moderate to high	Municipal wastewater upgrades	Meng *et al.* (2024)
Membrane bioreactors (MBR)	High biomass retention, membrane separation	COD: >95%; TSS: ~100%	High	Water reuse, space-limited plants	Jijingi *et al.* (2024)
Biofilm-based systems (MBBR, IFAS)	Attached growth on carrier media	COD: 80–90%; TN: 60–75%	Moderate	Variable-load wastewater	Andreottola *et al.* (2003)
Anaerobic/anammox systems	Low oxygen demand, shortcut nitrogen removal	COD: 70–90%; TN: >85%	Low	High-strength and side stream treatment	Cheng *et al.* (2025)

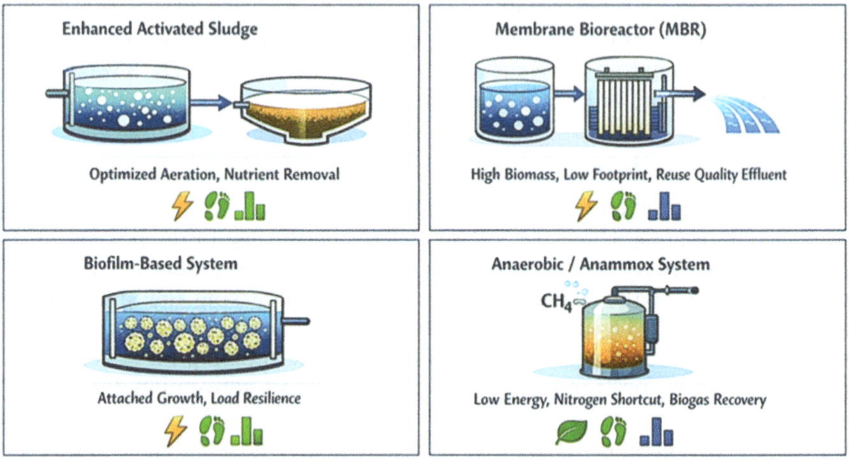

Figure 3.2 Advanced biological treatment configurations and operational characteristics.

3.3.1 Enhanced activated sludge systems

As described by Nagpal *et al.* (2024), improved activated sludge systems are a modification of traditional suspended growth processes and they include changes that are designed to make treatment more efficient and more resilient. The use of extended aeration, step-feeding, and biological nutrient removal systems have been designed to obtain more effective removal of organic matter, nitrogen, and phosphorus. These systems are designed to provide stable microbial communities and better process performance by optimizing aeration behavior, sludge age, and reactor sequence. A number of improvements have recently been added, including the addition of real-time monitoring tools and the automatic control of critical operational parameters such as dissolved oxygen and time to retain sludge. These measures reduce the sensitivity to fluctuation in influents and support consistent effluent quality. As a result, enhanced activated sludge systems are still widely deployed in facilities that are seeking to upgrade existing infrastructure without substantial structural change (P. Zhang *et al.*, 2025).

3.3.2 Membrane bioreactors and biofilm-based processes

As reported by Zulhendri *et al.* (2025), membrane bioreactors emulsify biological and physical separation with membrane filtration in place of conventional secondary clarification. This system enables high biomass concentrations and effective solid retention, resulting in improved effluent quality and shorter plant footprint. Membrane bioreactors are particularly suitable for extreme discharge requirements or water reuse, where the removal of suspended solids and pathogens is reliable. Moving-bed biofilm reactors and fixed-film systems are biofilm-based processes in which attached microbial growth provides treatment stability and loading capacity (Lariyah *et al.*, 2016). The retention of biomass on carrier media prevents washout and allows for different microbial populations to emerge. Because membrane systems also provide resistance to hydraulic and organic loads, biofilm systems are also attractive for handling complex or variable wastewater streams (Das & Mishra, 2025).

3.3.3 Anaerobic and anammox-based treatment systems

As highlighted by Zieliński *et al.* (2023), anaerobic treatment systems have gained renewed attention for their ability to treat heavy-lift wastewater in addition to producing biogas as a renewable energy source. Using processes such as anaerobic sludge blanket reactors and anaerobic membrane bioreactors, processes do not require aeration and as a result require much less energy. These systems are particularly effective in industrial wastewater and concentrated municipal streams, where organic levels are high enough to permit a stable anaerobic metabolism. Anammox-based treatment methods represent an important new form of biological nitrogen removal. As anammox processes are anoxic, they allow the direct conversion of ammonium and nitrite into nitrogen gas, which decreases the need for oxygen and carbon as compared to conventional nitrification–denitrification processes (Zhiji *et al.*, 2015). Although these systems require careful control and longer start-up periods, their energy efficiency and reduced sludge production make them increasingly attractive for mainstream and side stream treatment applications (Cho *et al.*, 2020).

3.4 INNOVATIVE CHEMICAL AND PHYSICOCHEMICAL TREATMENT APPROACHES

As per Muthukumar *et al.* (2021), the physicochemical and chemical treatment processes of pollutants are fundamental in the treatment of toxic biologically resistant pollutants. These methods can be particularly successful in eliminating reactive organic compounds, toxic industrial chemicals, color and odor, as well as new contaminants at very low levels (Nishat *et al.*, 2023). Recent innovation has been to improve reaction efficiency, reduce the chemical consumption, and combine these reactions with biological treatment to improve overall performance (Table 3.3). Unlike biological systems, chemical and physicochemical treatments can be conducted in the presence of well-defined reaction pathways and physical interactions that enable rapid pollutant transformation or separation. In combination with programmable monitoring and control systems, these technologies provide reliable, flexible solutions for multi-tier wastewater treatment applications (Toth *et al.*, 2022) (Figure 3.3).

3.4.1 Advanced oxidation processes

Advanced oxidation processes (AOPs) are designed to generate highly reactive oxidizing species such as •OH, with a strong non-selective oxidation potential at $E^0 \approx 2.8$ V. The radicals can mineralize a wide range of organic pollutants that cannot be removed by biological methods (Priyadarshini *et al.*, 2022).

Other common AOPs include ozonation, ultraviolet UV systems, Fenton and photo-Fenton reactions, and persulfate oxidation. The mechanism is the

Table 3.3 Comparison of chemical and physicochemical treatment technologies.

Technology	Primary Mechanism	Target Contaminants	Typical Performance	Key Limitations	References
AOPs	Hydroxyl or sulfate radical oxidation	Refractory organics, pharmaceuticals	80–95% organic removal	High energy demand, scavenging effects	Hübner *et al.* (2024)
Electrochemical treatment	Anodic oxidation and electro-generated oxidants	Industrial organics, color, metals	COD reduction: 60–90%	Electrode cost and fouling	Zhan *et al.* (2025)
Adsorption processes	Surface binding and pore diffusion	Dyes, trace organics, metals	Adsorption capacity: 50–500 mg/g	Regeneration and disposal issues	Jaber *et al.* (2024)
Hybrid chemical–biological systems	Partial oxidation followed by biodegradation	Complex industrial wastewater	Overall removal: >90%	Process integration complexity	Ghazal *et al.* (2022)

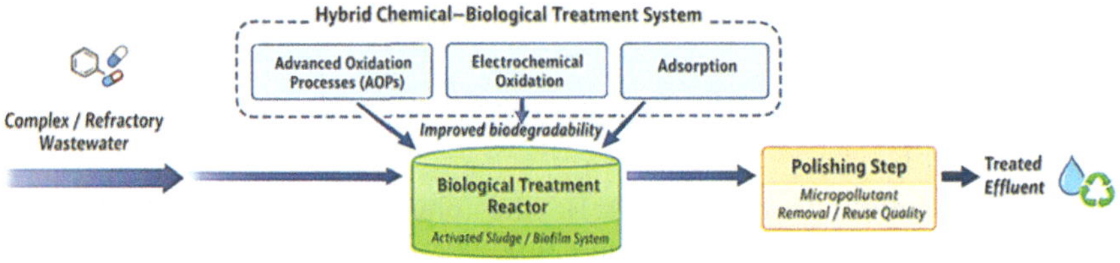

Figure 3.3 Integration of chemical, physicochemical, and hybrid treatment pathways.

in-situ generation of reactive radicals, as shown in equations below (Oturan & Aaron, 2014).

$$H_2O_2 + h\nu \rightarrow 2 \cdot OH$$
$$O_3 + H_2O \rightarrow 2 \cdot OH + O_2$$
$$Fe^{2+} + H_2O_2 \rightarrow Fe^{3+} + \cdot OH + OH^-$$

Once generated, hydroxyl radicals react rapidly with organic contaminants through hydrogen abstraction, electron transfer, or addition reactions, ultimately leading to their degradation into simpler compounds or complete mineralization:

$$\cdot OH + R \rightarrow CO_2 + H_2O + \text{inorganic ions}$$

Pharmaceuticals, dyes, and endocrine disrupting compounds have shown removal efficiencies that can reach 80–95% at optimized conditions with an average reaction time of minutes to tens of minutes (Nunes, 2022). However,

AOP performance is closely affected by bicarbonates, chlorides, and organic elements in the wastewater matrix that can act as radical scavengers. Thus, the use of catalyst optimization, hybrid AOP, and real-time dose control has led to studies addressing selectivity and energy efficiency through optimizing catalysts, establishing hybrid AOPs, and providing real-time dose control (Zheng *et al.*, 2025).

3.4.2 Electrochemical and adsorption-based technologies

As described by Viswanathan and Manisankar (2025), electrochemical treatments use electrical energy to control oxidation, reduction, or separation processes directly within the wastewater matrix. These include electrocoagulation, electrooxidation, and electro-Fenton processes which offer varying controllability and low chemical input.

In electrooxidation systems, contaminants are degraded either directly at the anode surface or indirectly via electro-generated oxidants such as hydroxyl radicals (Panizza & Cerisola, 2009), or active chlorine species:

$$H_2O \rightarrow \cdot OH + H^+ + e^-$$

$$Cl^- \rightarrow Cl_2 \rightarrow \frac{HOCl}{OCl^-}$$

Electrochemical processes have demonstrated chemical oxygen demand (COD) removal efficiencies between 60 and 90% for industrial wastewater streams and higher efficiencies when employing new electrode materials such as boron-doped diamond or mixed metal oxides. They are modular and easily automated, so they are well suited for small scale and decentralized treatment (Szpyrkowicz *et al.*, 2005).

Adsorption technologies are those that require the physical or chemical binding of pollutants onto the material. The active carbon, biochar, metal–organic frameworks, and functionalized nanomaterials have been widely investigated. The adsorption process is often described using isotherm models (Awad *et al.*, 2020), such as the Langmuir equation:

$$q_e = \frac{q_{\max} K_L C_e}{1 + K_L C_e}$$

where q_e is the amount adsorbed at equilibrium, C_e is the equilibrium concentration, $q_{\max}$ is the maximum adsorption capacity, and K_L is the Langmuir constant.

Reported adsorption capacities for dyes and pharmaceuticals range from 50 to over 500 mg/g, depending on material properties and surface functionalization. Challenges related to adsorbent regeneration and disposal have driven interest in integrating adsorption with catalytic or biological regeneration strategies.

3.4.3 Hybrid chemical–biological treatment systems

According to Tee *et al.* (2016), chemical–biological hybrid treatment systems combine the advantages of physicochemical and biological degradation to produce more efficacy and higher process resilience. Typically, in such systems, chemical or physicochemical pretreatment is used to transform complex or toxic compounds into more biodegradable intermediates that enhance the subsequent biological treatment performance. For example, partial oxidation using ozone or Fenton-based processes can increase the biodegradability index (BOD_5/COD ratio) of industrial wastewater from values below 0.2 to above 0.4, enabling effective downstream biological treatment (Gopalakrishnan *et al.*, 2023). Gavrilaş *et al.* (2026) describe, hybrid configurations have also been implemented in reverse order, where biological treatment is followed by targeted chemical polishing to remove residual micropollutants and pathogens. Such systems are increasingly applied in water reuse schemes, where consistent effluent quality is essential. In operational analysis, it has been shown that the combined systems have higher overall COD and nutrient removal efficiencies than the standard chemical treatment systems, and lower chemical demand and sludge production than the conventional systems. The integration of intelligent control systems further enhances performance by dynamically adjusting chemical dosing and process sequencing based on real-time biological activity and effluent quality (Tamburus *et al.*, 2025).

3.5 ARTIFICIAL INTELLIGENCE AND MACHINE LEARNING IN WASTEWATER TREATMENT

According to Baarimah *et al.* (2024), the use of artificial intelligence (AI) and machine learning (ML) in wastewater treatment has evolved due to the increasing complexity of treatment processes and growing availability of operational data. Modern treatment plants collect large quantities of information from online sensors, laboratory analysis, and supervisory control systems. While models of process are useful for designing and steady-state analysis, they often do not account for nonlinear interactions, temporal behavior, and site-specific variability that occur with full-scale operations (Lowe *et al.*, 2022). AI and ML approaches provide data-driven frameworks capable of learning patterns directly from historical and real-time datasets. Rather than replacing established process knowledge, these methods complement mechanistic understanding by improving prediction accuracy, supporting early fault detection, and enabling adaptive control strategies. Their value lies in translating complex data streams into actionable operational insights (Capodaglio & Callegari, 2025).

3.5.1 Data sources and model development

As emphasized by Zou *et al.* (2025) the quality and diversity of the available data is significant to the performance of AI and ML in wastewater treatment. Typical measurements from online sensors, such as flow rate, pH, dissolved oxygen, oxidation–reduction potential, turbidity and ammonia concentration, and laboratory-based measurements of biological oxygen demand (BOD), COD, total nitrogen, and phosphorus, are common sources of data (Liu *et al.*, 2025). Data from programmable logic controllers and maintenance history further enrich the inputs of models. Model development typically begins with data preprocessing, including outlier detection, noise filtering, normalization, and handling of missing values. Feature selection is critical, as redundant or weakly informative variables can degrade model performance. Widely used ML techniques include artificial neural networks (ANNs), support vector machines (SVMs), random forests, and, more recently, deep learning architectures for time-series analysis (Güçlü & Dursun, 2010).

A generic ANN-based predictive model can be expressed as

$$y = f\left(\sum_{i=1}^{n} w_i x_i + b\right)$$

where x_i are input variables (e.g., influent COD, flow rate, temperature), w_i are model weights, b is the bias term, and f represents a nonlinear activation function. Studies have shown that properly trained ML models can predict effluent COD, ammonia, or nitrate concentrations with coefficients of determination (R^2) exceeding 0.85–0.95 under stable operating conditions. Model generalization and transferability, however, remain important considerations, particularly when applying models across different treatment plants (Dalian *et al.*, 2026; Vlachou *et al.*, 2025).

3.5.2 Process monitoring and performance prediction

As per Andersen *et al.* (2024), integrated AI-based monitoring systems provide continuous monitoring of treatment outcomes wherein deviations from normal clinical practice are identified. ML classifiers and multivariate statistical techniques are commonly used to detect sensors drift, equipment failure, or process turbulence. Unlike the univariate threshold-based alarms, these systems examine relationships among multiple process variables, increasing sensitivity, and reducing false alarms (Ordieres-Meré *et al.*, 2025). Performance prediction is one of the most established applications of ML in wastewater treatment. Short-term forecasting of effluent quality parameters, such as ammonia or total nitrogen, allows operators to anticipate compliance risks and implement corrective actions proactively. Time-series models, including recurrent neural networks and long short-term memory networks, are particularly effective for capturing dynamic process behavior (El-Rawy *et al.*, 2021):

$$h_t = f(x_t, h_{t-1})$$

where h_t represents the hidden state at time t, incorporating both current inputs and historical process information. Reported prediction errors for effluent ammonia concentrations are often below 10–15% when sufficient training data are available. These predictive capabilities are especially valuable during periods of fluctuating influent loads or seasonal temperature changes (Farhi *et al.*, 2021).

3.5.3 Intelligent control and optimization strategies

According to D. Zhang *et al.* (2026), beyond monitoring and prediction, AI and ML support intelligent control strategies that aim to optimize process performance while minimizing energy and chemical consumption. Model predictive control (MPC) frameworks increasingly incorporate ML-based models to improve prediction accuracy in nonlinear systems. In such systems, future control actions are selected by minimizing an objective function subject to operational constraints:

$$\min_{u(t)} \sum_{k=1}^{N} \left[(y_k - y_{ref})^2 + \lambda u_k^2 \right]$$

where y_k is the predicted process output, y_{ref} is the desired setpoint, u_k represents control actions (e.g., aeration rate), and λ is a weighting factor.

Applications of intelligent control include adaptive aeration control, dynamic chemical dosing, and optimized sludge retention time management. Full-scale implementations have reported energy savings of 10–25% in aeration systems while maintaining or improving effluent quality. Importantly, these strategies rely on transparent model validation and operator oversight to ensure reliability and regulatory compliance (Philips *et al.*, 2025).

The integration of AI-based optimization with existing control architectures allows treatment plants to transition toward more adaptive operation without compromising system robustness. As experience with these tools grows, their role is likely to expand from decision support to semi-autonomous control within clearly defined operational boundaries (Yang *et al.*, 2025).

3.6 AUTOMATION AND SMART PROCESS CONTROL

As noted by Olsson (2006), modern wastewater treatment plants are powered by automation and smart process control. The complexity of treatment systems and increasing expectations of performance has made manual control and periodic monitoring no longer an option to assure smooth and efficient operation. Automation allows for continuous process monitoring, rapid response to disruptions and frequent control practices in varying operating conditions (Krishnan *et al.*, 2022) (Figure 3.4).

As reported by Mansoor *et al.* (2025), smart process control extends conventional automation by integrating advanced sensing, data analytics, and adaptive control algorithms. Rather than relying solely on fixed setpoints, these systems interpret real-time process data to adjust operational parameters in a manner that reflects current treatment demands. This shift supports improved effluent quality, reduced resource consumption, and enhanced operational reliability.

3.6.1 Sensor integration and real-time system feedback

According to Narayana *et al.* (2024), smart process control starts with the use of reliable sensors that can provide a continuous and high-resolution data set on key process variables. Typical monitored parameters include flow rate, pH, dissolved oxygen, temperature, turbidity, oxidation–reduction potential, ammonia, nitrate, and phosphate concentrations. Sensor advances have improved measurement stability, lessen maintenance requirements and made online monitoring in very harsh wastewater environments more practicable.

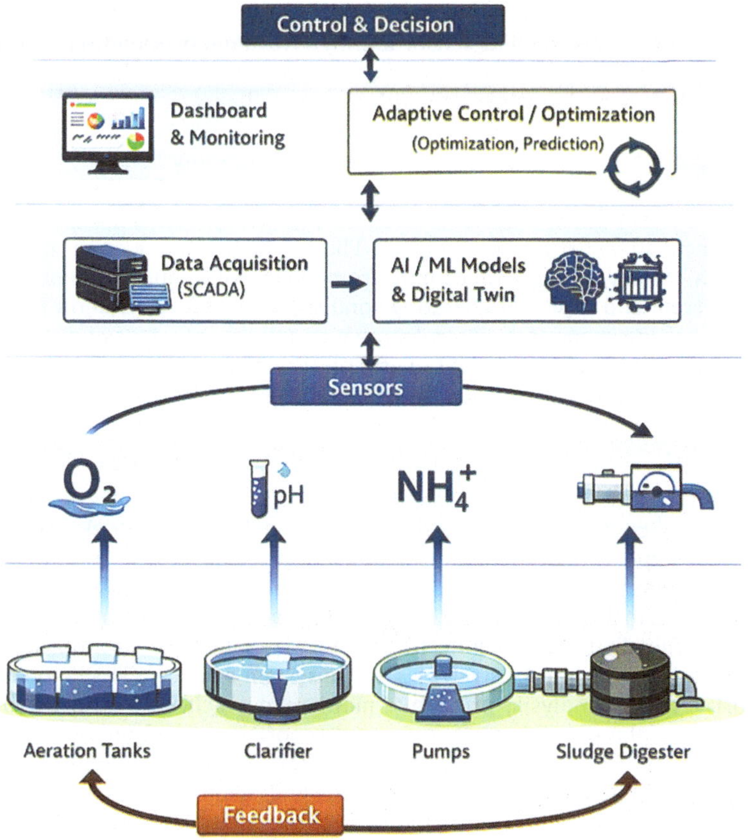

Figure 3.4 Smart process control architecture for intelligent wastewater treatment.

Sensor data are transmitted to supervisory control and data-acquisition systems, where real-time feedback loops support automated control actions. For example, aeration control systems adjust oxygen supply based on measured dissolved oxygen and ammonia concentrations to maintain optimal biological activity:

$$u(t) = K_p e(t) + K_i \int e(t)\, dt + K_d \frac{de(t)}{dt}$$

where $u(t)$ represents the control input, $e(t)$ is the deviation from the target setpoint, and K_p, K_i, and K_d are controller gains. While proportional–integral–derivative control remains widely used, its effectiveness is enhanced when combined with real-time data validation and adaptive tuning strategies (Parsa *et al.*, 2024).

Real-time feedback systems also support early detection of abnormal operating conditions, such as sensor fouling, hydraulic overloading, or process imbalance. By enabling timely intervention, automation reduces the risk of treatment failure and improves overall system resilience.

3.6.2 Digital twins and adaptive operational frameworks

Wang *et al.* (2024) reported, the development of digital twins, the virtual equivalent to real-life wastewater treatment plants, represents a significant new development in smart process control. These models take account of process equations, history, and real-time inputs from sensors to simulate system behavior under current and expected operating conditions. Digital twins allow operators to assess process responses, control strategies and the influence of disturbances without affecting real-world plant operation (Rodríguez-Alonso *et al.*, 2024).

A simplified digital twin framework can be represented as

$$\hat{y}(t) = f(x(t), u(t), \theta)$$

where $\hat{y}(t)$ is the predicted system output, $x(t)$ represents process states, $u(t)$ denotes control actions, and θ includes calibrated model parameters. Continuous updating of the digital twin using real-time data allows the virtual model to remain aligned with the physical system (Tao *et al.*, 2019).

According to Khan *et al.* (2025), the adaptive operational frameworks are based on the digital twin concept and incorporate both optimization and learning to change control strategies as the system evolves. These models also support scenario analysis, predictive maintenance scheduling, and long-run performance planning. Applications cited report increased energy efficiency, shortening of downtime, and more consistent compliance with discharge requirements. The adoption of digital twins and adaptive control frameworks requires careful consideration of model accuracy, data integrity, and operator trust. When implemented with appropriate validation and oversight, these tools provide a practical pathway toward more responsive and efficient wastewater treatment operations (Ogunnowo *et al.*, 2024; Subeesh & Chauhan, 2026).

3.7 IMPLEMENTATION CHALLENGES AND PRACTICAL CONSIDERATIONS

While modern and intelligent wastewater treatment technologies do offer clear performance and efficiency advantages, the application of these technologies entails multiple technical, economic, and institutional challenges. Many treatment facilities are hampered by current infrastructure, regulatory limitations and limited operational flexibility. As a result, the transition toward intelligent treatment systems often occurs incrementally rather than through complete system replacement (Ejairu *et al.*, 2024; Sha *et al.*, 2024) (Table 3.4).

Successful implementation depends not only on technological capability but also on alignment with site-specific conditions, operator expertise, and long-term operational objectives. Addressing these practical considerations is essential to ensure that advanced technologies deliver measurable and

Table 3.4 Applications and benefits of intelligent control strategies in wastewater treatment.

Intelligent Application	Control Objective	Operational Benefit	Reported Impact	References
AI-based effluent prediction	Anticipate quality deviations	Early corrective actions	Prediction accuracy (R^2): 0.85–0.95	Ali and Ahmed (2024)
Adaptive aeration control	Match oxygen supply to demand	Reduced energy consumption	Energy savings: 10–25%	Jamaludin *et al.* (2024)
Digital twin-assisted operation	Scenario testing and optimization	Improved operational confidence	Reduced downtime and disturbances	Mingorance *et al.* (2025)
Automated chemical dosing	Minimize overdosing	Cost and sludge reduction	Chemical savings: 15–30%	Cairone *et al.* (2024)

sustained benefits under real-world operating conditions (Ghobakhloo *et al.*, 2026; Jamil *et al.*, 2025).

3.7.1 System scalability and retrofitting existing plants

One of the primary challenges in deploying advanced treatment technologies is scalability, particularly when transitioning from pilot-scale demonstrations to full-scale plant operation. Technologies that perform well under controlled conditions may exhibit different behavior when exposed to variable influent loads, equipment limitations, and complex hydraulic interactions in existing facilities. Scaling decisions must therefore account for process integration, spatial constraints, and compatibility with legacy systems (Gavrilaș *et al.*, 2026; Meng *et al.*, 2024).

Suprun *et al.* (2022) find that the, retrofitting existing wastewater treatment plants presents additional challenges. Many facilities were designed with limited space and rigid process layouts, restricting the installation of new reactors, sensors, or automation infrastructure. Integrating advanced biological, chemical, or digital components often requires phased implementation to avoid service disruptions. For example, the addition of online sensors or advanced control systems may necessitate upgrades to electrical systems, data networks, and supervisory control platforms (Ignatkina *et al.*, 2025).

Economic considerations also influence scalability decisions. Capital investment requirements, lifecycle costs, and uncertainty in performance outcomes can limit adoption, particularly for smaller utilities. Modular and decentralized solutions have gained attention as scalable alternatives, allowing utilities to incrementally expand capacity or functionality in response to changing treatment demands (Duran-Romero & Barquet, 2025; Rahman *et al.*, 2020).

3.7.2 Operational reliability and workforce requirements

As noted by Lee and Singh (2020), when introducing new, intelligent treatment technologies, reliability of operation remains an important concern. The growing complexity of the system can introduce new failure points such as sensor failure, data communication problems, or software failures. High-quality systems and components, redundancy and manual override for abnormal operation are key to maintaining stable operation.

As highlighted by Huzooree *et al.* (2025), workforce readiness is equally important. Advanced treatment systems demand skills that extend beyond traditional plant operation, including data interpretation, basic model understanding, and familiarity with automated control interfaces. In many cases, successful implementation depends on targeted training programs that bridge the gap between conventional operational expertise and digital system management (Rožman *et al.*, 2023).

Institutional factors such as staff turnover, resistance to technological change, and limited technical support can further affect system performance. Ensuring that operators remain actively engaged in decision-making, rather than relegated to passive system monitoring, is essential for building trust in intelligent control systems. When workforce development is addressed alongside technological deployment, advanced treatment systems are more likely to achieve long-term operational stability and acceptance (Petropoulos *et al.*, 2026).

3.8 FUTURE DIRECTIONS IN INTELLIGENT WASTEWATER TREATMENT

The continued evolution of wastewater treatment systems represents the abrader shift to an infrastructure that is not only functional but also adaptive to uncertainty and change. Process engineering, sensing, and data analytics have begun to reshape treatment plant design and operation. Future smart wastewater treatment projects should aim to improve the adaptability of the system, ensuring better integrated treatment across stages, and strengthening the relationship between technology and process decision-making. Rather than pursuing technological sophistication as an end in itself, future progress will depend on addressing unresolved challenges related to robustness, transparency, and long-term performance. Research and innovation efforts must therefore balance ambition with practical applicability to ensure that intelligent systems contribute meaningfully to sustainable wastewater management (Garrido-Baserba *et al.*, 2024; Murat Hocaoğlu *et al.*, 2025).

3.8.1 Emerging technologies and research needs

Several emerging technologies are likely to influence the next generation of intelligent wastewater treatment systems. Advances in sensor technology, including low-maintenance nutrient sensors and spectroscopic methods for real-time water quality assessment, are expanding the scope of online monitoring. Improved data resolution and reliability will support more accurate process modelling and control, particularly for complex biological and hybrid treatment systems. On the computational side, continued development of hybrid modelling approaches that combine mechanistic process models with data-driven learning techniques represents an important research direction. Such models can leverage established process understanding while accommodating site-specific variability and nonlinear behavior. Research is also needed to improve model transferability across different treatment plants and operating conditions, reducing the need for extensive retraining. Additional research priorities include quantifying uncertainty in AI-based predictions, developing standardized validation protocols, and assessing the long-term stability of intelligent control systems under real operating conditions. Addressing these gaps will be essential for building confidence among regulators, operators, and utility managers (Chen *et al.*, 2025; Mythili *et al.*, 2026).

3.8.2 Toward fully adaptive and autonomous treatment systems

The concept of fully adaptive and autonomous wastewater treatment systems is gaining attention as digital technologies mature. In such systems, treatment processes would continuously adjust to influent characteristics, operational constraints, and performance objectives with minimal manual intervention. Automation, predictive modelling, and optimization would operate within predefined safety and regulatory boundaries, allowing human operators to focus on oversight and strategic decision-making. Progress toward autonomy is expected to occur incrementally. Semi-autonomous systems that support operator decisions and automate routine control tasks are already being implemented in many facilities. Further advances will require improvements in system reliability, fail-safe design, and transparency of control logic to ensure that automated decisions remain understandable and traceable. Equally important are governance and institutional considerations. Clear accountability frameworks, regulatory acceptance, and workforce adaptation will shape how far autonomy can be realized in practice. When developed responsibly, adaptive and autonomous systems have the potential to enhance treatment resilience, reduce operational burden, and support long-term sustainability goals without displacing the critical role of human expertise.

3.9 CONCLUSION

Developing wastewater treatment technologies also reflects a larger understanding that traditional design of processes alone cannot adequately meet the increasing complexity, variability, and performance requirements of modern wastewater treatment. As regulatory requirements become more stringent and sustainability goals become important, treatment systems must function more efficiently, more resiliently, and more adaptably than before. The progress of traditional treatments through advanced biological, chemical, and physicochemical processes has been discussed in this chapter, where AI, automation, and smart process control are playing a growing role. Intelligent technologies are not a deviation from accepted treatment practices; they replace existing systems by improving process understanding, proactive decision making, and adaptive operation in changing environments. Integrated data models, real-time monitoring, and optimization strategies achieve tangible impacts in terms of effluent quality, energy efficiency, and operation stability. Nonetheless, practical challenges in scalability, system integration, operational reliability, and workforce readiness remain a core component of successful implementation. To address these concerns, it is necessary to balance technological innovation with site-specific constraints and institutional capacity. The most successful approaches will likely combine solid engineering and well-defined operating systems, thus facilitating sustainable and reliable treatment systems that can fully satisfy present and future conditions, as wastewater treatment has evolved.

REFERENCES

Abdelrahman A. M., Kosar S., Gulhan H., Cicekalan B., Ucas G., Atli E., Guven H., Ozgun H., Ozturk I., Koyuncu I. and van Lier J. B. (2023). Impact of primary treatment methods on sludge characteristics and digestibility, and wastewater treatment plant-wide economics. *Water Research*, **235**, 119920. https://doi.org/10.1016/j.watres.2023.119920

Ali A. J. and Ahmed A. A. (2024). Long-term AI prediction of ammonium levels in rivers using transformer and ensemble models. *Cleaner Water*, **2**, 100051. https://doi.org/10.1016/j.clwat.2024.100051

Andersen E. S., Birk-Korch J. B., Hansen R. S., Fly L. H., Röttger R., Cespedes Arcani D. M., Brasen C. L., Brandslund I. and Madsen J. S. (2024). Monitoring performance of clinical artificial intelligence in health care: a scoping review. *JBI Evidence Synthesis*, **22**(12), 2423–2446. https://doi.org/10.11124/JBIES-24-00042

Andreottola G., Foladori P., Gatti G., Nardelli P., Pettena M. and Ragazzi M. (2003). Upgrading of a small overloaded activated sludge plant using a MBBR system. *Journal of Environmental Science and Health. Part A, Toxic/Hazardous Substances & Environmental Engineering*, **38**, 2317–2328. https://doi.org/10.1081/ESE-120023388

Awad A. M., Jalab R., Benamor A., Nasser M. S., Ba-Abbad M. M., El-Naas M. and Mohammad A. W. (2020). Adsorption of organic pollutants by nanomaterial-based adsorbents: an overview. *Journal of Molecular Liquids*, **301**, 112335. https://doi.org/10.1016/j.molliq.2019.112335

Baarimah A. O., Bazel M. A., Alaloul W. S., Alazaiza M. Y. D., Al-Zghoul T. M., Almuhaya B., Khan A. and Mushtaha A. W. (2024). Artificial intelligence in wastewater treatment: research trends and future perspectives through bibliometric analysis. *Case Studies in Chemical and Environmental Engineering*, **10**, 100926. https://doi.org/10.1016/j.cscee.2024.100926

Cairone S., Hasan S. W., Choo K.-H., Lekkas D. F., Fortunato L., Zorpas A. A., Korshin G., Zarra T., Belgiorno V. and Naddeo V. (2024). Revolutionizing wastewater treatment toward circular economy and carbon neutrality goals: pioneering sustainable and efficient solutions for automation and advanced process control with smart and cutting-edge technologies. *Journal of Water Process Engineering*, **63**, 105486. https://doi.org/10.1016/j.jwpe.2024.105486

Capodaglio A. G. and Callegari A. (2025). Use, potential, needs, and limits of AI in wastewater treatment applications. *Water*, **17**(2), 170. https://doi.org/10.3390/w17020170

Chen H., Gao X. and Yuan R. (2025). Advances in remote sensing and sensor technologies for water-quality monitoring: a review. *Water*, **17**(20), 3000. https://doi.org/10.3390/w17203000

Cheng Q., Chen Y., Zhao W., Guo X., Nengzi L., Gong Z., Du E. and Peng M. (2025). Anammox enhanced anoxic/oxic/oxic/anoxic process for removing carbon, nitrogen, and phosphorus from extremely low carbon/nitrogen ratio wastewater: feasibility, performance and mechanism. *Bioresource Technology*, **440**, 133539. https://doi.org/10.1016/j.biortech.2025.133539

Cho S., Kambey C. and Nguyen V. K. (2020). Performance of anammox processes for wastewater treatment: a critical review on effects of operational conditions and environmental stresses. *Water*, **12**(1), 0. https://doi.org/10.3390/w12010020

Dalian Y., Junjun Z. and Hui L. (2026). Capsule networks for intelligent fault diagnosis: a roadmap of recent advancements and challenges. *Expert Systems with Applications*, **296**, 128814. https://doi.org/10.1016/j.eswa.2025.128814

Das A. and Mishra S. (2025). Reimagining biofiltration for sustainable industrial wastewater treatment. *Discover Sustainability*, **6**(1), 826. https://doi.org/10.1007/s43621-025-01784-8

Duran-Romero D. and Barquet K. (2025). Business models for decentralized water services in urban and peri-urban areas. *Cleaner Water*, **4**, 100138. https://doi.org/10.1016/j.clwat.2025.100138

Ejairu U., Aderamo A., Olisakwe H., Esiri A., Uwaga M. and Solomon N. (2024). Eco-friendly wastewater treatment technologies (concept): conceptualizing advanced, sustainable wastewater treatment designs for industrial and municipal applications. *Comprehensive Research and Reviews in Engineering and Technology*, **2**, 83–104. https://doi.org/10.57219/crret.2024.2.1.0063

El-Rawy M., Abd-Ellah M., Fathi H. and Ahmed A. (2021). Forecasting effluent and performance of wastewater treatment plant using different machine learning techniques. *Journal of Water Process Engineering*, **44**(October), 1–15. https://doi.org/10.1016/j.jwpe.2021.102380

Essamlali I., Nhaila H. and Khaili M. E. (2024). Advances in machine learning and IoT for water quality monitoring: a comprehensive review. *Heliyon*, **10**(6), e27920. https://doi.org/10.1016/j.heliyon.2024.e27920

Farhi N., Kohen E., Mamane H. and Shavitt Y. (2021). Prediction of wastewater treatment quality using LSTM neural network. *Environmental Technology & Innovation*, **23**, 101632. https://doi.org/10.1016/j.eti.2021.101632

Fernandes J., Ramísio P. J. and Puga H. (2024). A comprehensive review on various phases of wastewater technologies: trends and future perspectives. *Eng*, **5**(4), 2633–2661. https://doi.org/10.3390/eng5040138

Gad M., Abdo S. M., Hu A., El-Liethy M. A., Hellal M. S., Doma H. S. and Ali G. H. (2022). Performance assessment of natural wastewater treatment plants by multivariate statistical models: a case study. *Sustainability*, **14**(13), 7658. https://doi.org/10.3390/su14137658

Garrido-Baserba M., Sedlak D. L., Molinos-Senante M., Barnosell I., Schraa O., Rosso D., Verdaguer M. and Poch M. (2024). Using water and wastewater decentralization to enhance the resilience and sustainability of cities. *Nature Water*, **2**(10), 953–974. https://doi.org/10.1038/s44221-024-00303-9

Gavrilaş S., Gerőcs T., Chereji B.-D. and Munteanu F.-D. (2026). Trends in advanced wastewater treatment technologies: from membrane bioreactors to advanced oxidation processes. *Water*, **18**(3), 350. https://doi.org/10.3390/w18030350

Ghazal H., Koumaki E., Hoslett J., Malamis S., Katsou E., Barcelo D. and Jouhara H. (2022). Insights into current physical, chemical and hybrid technologies used for the treatment of wastewater contaminated with pharmaceuticals. *Journal of Cleaner Production*, **361**, 132079. https://doi.org/10.1016/j.jclepro.2022.132079

Ghimire U., Sarpong G. and Gude V. G. (2021). Transitioning wastewater treatment plants toward circular economy and energy sustainability. *ACS Omega*, **6**(18), 11794–11803. https://doi.org/10.1021/acsomega.0c05827

Ghobakhloo M., Fathi M., Okwir S., Al-Emran M. and Ivanov D. (2026). Adaptive social manufacturing: A human-centric, resilient, and sustainable framework for advancing industry 5.0. *International Journal of Production Research*, **64**(3), 1127–1160. https://doi.org/10.1080/00207543.2025.2559137

Gopalakrishnan G., Jeyakumar R. B. and Somanathan A. (2023). Challenges and emerging trends in advanced oxidation technologies and integration of advanced oxidation processes with biological processes for wastewater treatment. *Sustainability*, **15**(5), 4235. https://doi.org/10.3390/su15054235

Gu Y., Li Y., Yuan F. and Yang Q. (2023). Optimization and control strategies of aeration in WWTPs: a review. *Journal of Cleaner Production*, **418**, 138008. https://doi.org/10.1016/j.jclepro.2023.138008

Güçlü D. and Dursun S. (2010). Artificial neural network modelling of a large-scale wastewater treatment plant operation. *Bioprocess and Biosystems Engineering*, **33**(9), 1051–1058. https://doi.org/10.1007/s00449-010-0430-x

Gupta A. S. and Khatiwada D. (2024). Investigating the sustainability of biogas recovery systems in wastewater treatment plants-–A circular bioeconomy approach. *Renewable and Sustainable Energy Reviews*, **199**, 114447. https://doi.org/10.1016/j.rser.2024.114447

Hübner U., Spahr S., Lutze H., Wieland A., Rüting S., Gernjak W. and Wenk J. (2024). Advanced oxidation processes for water and wastewater treatment–Guidance for systematic future research. *Heliyon*, **10**(9), e30402. https://doi.org/10.1016/j.heliyon.2024.e30402

Huzooree G., Subramanian J. and Dewasiri N. J. (2025). Reskilling for the AI age: Project management approaches and organisational strategies for workforce readiness. In M. Yadav, A. Pandey and G. Huzooree (eds), *Global Work Arrangements and Outsourcing in the Age of AI*, pp. 471–494. IGI Global Scientific Publishing. Doi: 10.4018/979-8-3373-1270-5

Ignatkina D., Yuriev Y., Bikunova M., Vlasova O., Tekushin D. and Gudeeva D. (2025). Practical experience in retrofitting existing water treatment plants due to stricter regulatory requirements for drinking water quality. *ВЕСТНИК ВОЛГОГРАДСКОГО ГОСУДАРСТВЕННОГО АРХИТЕКТУРНО-СТРОИТЕЛЬНОГО УНИВЕРСИТЕТА*, **January**, 127–139. https://doi.org/10.35211/18154360_2025_2_127

Jaber L., Backer S. N., Laoui T., Abumadi F., Saaid Koujan M. M., Khalil K. A., Shanableh A. and Atieh M. A. (2024). Recent trends in surface impregnation techniques on activated carbon for efficient pollutant removal from wastewater. *Desalination and Water Treatment*, **319**, 100562. https://doi.org/10.1016/j.dwt.2024.100562

Jain H. and Shah M. P. (2025). Microbial Insights Into Wastewater Treatment and Environmental Sustainability. Bentham Science Publishers, Sharjah, United Arab Emirates.

Jamaludin M., Tsai Y.-C., Lin H.-T., Huang C.-Y., Choi W., Chen J.-G. and Sean W.-Y. (2024). Modeling and control strategies for energy management in a wastewater center: a review on aeration. *Energies*, **17**(13), 3162. https://doi.org/10.3390/en17133162

Jamil K., Zhang W., Anwar A. and Mustafa S. (2025). Exploring the influence of AI adoption and technological readiness on sustainable performance in Pakistani export sector manufacturing small and medium-sized enterprises. *Sustainability*, **17**(8), 3599. https://doi.org/10.3390/su17083599

Jijingi H. E., Yazdi S. K., Abakar Y. A. and Etim E. (2024). Evaluation of membrane bioreactor (MBR) technology for industrial wastewater treatment and its application in developing countries: a review. *Case Studies in Chemical and Environmental Engineering*, **10**, 100886. https://doi.org/10.1016/j.cscee.2024.100886

Kabir T. & Anick K. M. T. A. (2024). PLC–SCADA–Integrated Electrical Automation Frameworks for Process Optimization in Water and Wastewater Treatment Facilities. *Review of Applied Science and Technology*, **3**(1), 221–262.

Khan T., Khan U., Khan A., Mollan C., Morkvenaite-Vilkonciene I. and Pandey V. (2025). Data-driven digital twin framework for predictive maintenance of smart manufacturing systems. *Machines*, **13**(6), 481. https://doi.org/10.3390/machines13060481

Kolosovska T. and Bauer S. K. (2022). Evaluation of grit properties at a medium-capacity wastewater treatment plant: a case study. *Resources Conservation & Recycling Advances*, **14**, 200080. https://doi.org/10.1016/j.rcradv.2022.200080

Krishnan S. R., Nallakaruppan M. K., Chengoden R., Koppu S., Iyapparaja M., Sadhasivam J. and Sethuraman S. (2022). Smart water resource management using artificial intelligence—A review. *Sustainability*, **14**(20), 13384. https://doi.org/10.3390/su142013384

Lariyah M. S., Mohiyaden H. A., Hayder G., Hayder G., Hussein A., Basri H., Sabri A. F. and Noh M. N. (2016). Application of moving bed biofilm reactor (MBBR) and integrated fixed activated sludge (IFAS) for biological river water purification system: a short review. *IOP Conference Series: Earth and Environmental Science*, **32**(1), 12005. https://doi.org/10.1088/1755-1315/32/1/012005

Lee J., Singh J., Ni J., Jiang B., Azamfar M. and Feng J. (2020). Intelligent Maintenance Systems and Predictive Manufacturing, *Journal of Manufacturing Science and Engineering* **142**(11), 110805, https://doi.org/10.1115/1.4047856

Liu Y., Mancuso G., Petrotto L., Lavrnić S., Dong Z., Tian Y., Zhang J. and Toscano A. (2025). AI-driven solutions in wastewater treatment and agricultural reuse systems: a comprehensive review. *Journal of Environmental Management*, **393**, 127008. https://doi.org/10.1016/j.jenvman.2025.127008

Lowe M., Qin R. and Mao X. (2022). A review on machine learning, artificial intelligence, and smart technology in water treatment and monitoring. *Water*, **14**(9), 1384. https://doi.org/10.3390/w14091384

Mansoor S., Iqbal S., Popescu S. M., Kim S. L., Chung Y. S. and Baek J.-H. (2025). Integration of smart sensors and IOT in precision agriculture: trends, challenges and future prospectives. *Frontiers in Plant Science*, **16**, 1587869. https://doi.org/10.3389/fpls.2025.1587869

Meng X., Huang Z. and Ge G. (2024). Upgrade and reconstruction of biological processes in municipal wastewater treatment plants. *Desalination and Water Treatment*, **317**, 100299. https://doi.org/10.1016/j.dwt.2024.100299

Mingorance R., Pereira D. C., Uribetxebarria J. and Leturiondo U. (2025). A methodology leveraging digital twins to enhance the operational strategy of manufacturing plants in unexpected scenarios. *Results in Engineering*, **27**, 106761. https://doi.org/10.1016/j.rineng.2025.106761

Moretti A., Ivan H. L. and Skvaril J. (2024). A review of the state-of-the-art wastewater quality characterization and measurement technologies. is the shift to real-time monitoring nowadays feasible? *Journal of Water Process Engineering*, **60**, 105061. https://doi.org/10.1016/j.jwpe.2024.105061

Murat Hocaoğlu S., Roghani B., Gülcan H., Magna D., Aydöner C., Barros V., Aynur S. K., Ratnaweera H., Fatone F., Eusebi A. and Maletskyi Z. (2025). Artificial intelligence in wastewater treatment plants: a review of current trends and adaptation strategies in the face of climate-driven rainfall fluctuations. *Journal of Water and Climate Change*, **16**(9), 2742–2759. https://doi.org/10.2166/wcc.2025.849

Muthukumar H., Shanmugam M. K., Dash S. S. and Gummadi S. N. (2021). Comparison of biological and physicochemical techniques for treatment of coffee wastewater—A comprehensive review. In: Biological Treatment of Industrial Wastewater, M. P.

Shah (ed.), The Royal Society of Chemistry, pp. 391–409. https://doi.org/10.1039/9781839165399
Mythili S., Neela K. L., Mahesh Selvi T. and Sikkandar M. Y. (2026). Smart IoT–ML–RS framework for detecting emerging wastewater contaminants. *Desalination and Water Treatment*, **325**, 101621. https://doi.org/10.1016/j.dwt.2025.101621
Nagpal M., Siddique M. A., Sharma K., Sharma N. and Mittal A. (2024). Optimizing wastewater treatment through artificial intelligence: recent advances and future prospects. *Water Science and Technology*, **90**(3), 731–757. https://doi.org/10.2166/wst.2024.259
Narayana T. L., Venkatesh C., Kiran A., Chinna Babu J., Kumar A., Khan S. B., Almusharraf A. and Quasim M. T. (2024). Advances in real time smart monitoring of environmental parameters using IoT and sensors. *Heliyon*, **10**(7), e28195. https://doi.org/10.1016/j.heliyon.2024.e28195
Nishat A., Yusuf M., Qadir A., Ezaier Y., Vambol V., Ijaz Khan M., Moussa S. B., Kamyab H., Sehgal S. S., Prakash C. and Yang H. H. (2023). Wastewater treatment: a short assessment on available techniques. *Alexandria Engineering Journal*, **76**, 505–516. https://doi.org/10.1016/j.aej.2023.06.054
Nishmitha P. S., Akhilghosh K. A., Aiswriya V. P., Ramesh A., Muthuchamy M. and Muthukumar A. (2025). Understanding emerging contaminants in water and wastewater: a comprehensive review on detection, impacts, and solutions. *Journal of Hazardous Materials Advances*, **18**, 100755. https://doi.org/10.1016/j.hazadv.2025.100755
Nunes R. (2022). Degradation of Contaminants with Different Chemical Structures by Reactive Oxygen and Oxysulfur Species Generated via Sulfite and Persulfate-Based Processes, unpublished doctoral thesis, University of Sao Paolo, Brazil.
Obaideen K., Shehata N., Sayed E. T., Abdelkareem M. A., Mahmoud M. S. and Olabi A. G. (2022). The role of wastewater treatment in achieving sustainable development goals (SDGs) and sustainability guideline. *Energy Nexus*, **7**, 100112. https://doi.org/10.1016/j.nexus.2022.100112
Ogunnowo E., Adewoyin M., Fiemotongha J., Igunma T. and Adeleke A. (2024). A conceptual framework for digital twin deployment in real-time monitoring of mechanical systems. *International Journal of Advanced Multidisciplinary Research and Studies*, **4**, 2350–2367. https://doi.org/10.62225/2583049X.2024.4.6.4353
Olsson G. (2006). Instrumentation, control and automation in the water industry–State-of-the-art and new challenges. *Water Science and Technology: A Journal of the International Association on Water Pollution Research*, **53**, 1–16. https://doi.org/10.2166/wst.2006.097
Ordieres-Meré J., Sánchez-Herguedas A. and Mena-Nieto Á. (2025). A data-driven monitoring system for a prescriptive maintenance approach: supporting reinforcement learning strategies. *Applied Sciences*, **15**(12), 6917. https://doi.org/10.3390/app15126917
Oturan M. and Aaron J.-J. (2014). Advanced oxidation processes in water/wastewater treatment: principles and applications. A review. *Critical Reviews in Environmental Science and Technology*, **44**(23), 2577–2641. https://doi.org/10.1080/10643389.2013.829765
Palermo S. A., Maiolo M., Brusco A. C., Turco M., Pirouz B., Greco E., Spezzano G. and Piro P. (2022). Smart technologies for water resource management: an overview. *Sensors*, **22**, 6225. https://doi.org/10.3390/s22166225
Panizza M. and Cerisola G. (2009). Direct and mediated anodic oxidation of organic pollutants. *Chemical Reviews*, **109**(12), 6541–6569. https://doi.org/10.1021/cr9001319
Parsa Z., Dhib R. and Mehrvar M. (2024). Dynamic modelling, process control, and monitoring of selected biological and advanced oxidation processes for wastewater

treatment: a review of recent developments. *Bioengineering*, **11**(2), 189. https://doi.org/10.3390/bioengineering11020189

Petropoulos F., Akkermans H., Zeynep Aksin O., Ali I., Babai M. Z., Barbosa-Povoa A., Battaïa O., Besiou M., Boysen N., Brammer S. and Brandon-Jones A. (2026). Operations & supply chain management: principles and practice. *International Journal of Production Research*, **64**(1), 330–513. https://doi.org/10.1080/00207543.2025.2555531

Philips B., Charles C. and Adebayo Y. (2025). Advanced Control Strategy: Model Predictive Control (MPC).

Pophali G. R., Kaul S. N. and Mathur S. (2003). Influence of hydraulic shock loads and TDS on the performance of large-scale CETPs treating textile effluents in India. *Water Research*, **37**(2), 353–361. https://doi.org/10.1016/S0043-1354(02)00268-3

Pratap B., Kumar S., Nand S., Azad I., Bharagava R. N., Romanholo Ferreira L. F. and Dutta V. (2023). Wastewater generation and treatment by various eco-friendly technologies: possible health hazards and further reuse for environmental safety. *Chemosphere*, **313**, 137547. https://doi.org/10.1016/j.chemosphere.2022.137547

Priyadarshini M., Das I., Ghangrekar M. M. and Blaney L. (2022). Advanced oxidation processes: performance, advantages, and scale-up of emerging technologies. *Journal of Environmental Management*, **316**, 115295. https://doi.org/10.1016/j.jenvman.2022.115295

Raabe D. (2023). The materials science behind sustainable metals and alloys. *Chemical Reviews*, **123**(5), 2436–2608. https://doi.org/10.1021/acs.chemrev.2c00799

Rahman M., Oni A., Gemechu E. and Kumar A. (2020). The development of techno-economic models for the assessment of utility-scale electro-chemical battery storage systems. *Applied Energy*, **283**, 1–15. https://doi.org/10.1016/j.apenergy.2020.116343

Rodríguez-Alonso C., Pena-Regueiro I. and García Ó. (2024). Digital twin platform for water treatment plants using microservices architecture. *Sensors*, **24**(5). https://doi.org/10.3390/s24051568

Rožman M., Oreški D. and Tominc P. (2023). Artificial-intelligence-supported reduction of employees' workload to increase the company's performance in today's VUCA environment. *Sustainability*, **15**(6), 5019. https://doi.org/10.3390/su15065019

Satyam S. and Patra S. (2024). Innovations and challenges in adsorption-based wastewater remediation: a comprehensive review. *Heliyon*, **10**(9), e29573. https://doi.org/10.1016/j.heliyon.2024.e29573

Sha C., Shen S., Zhang J., Zhou C., Lu X. and Zhang H. (2024). A review of strategies and technologies for sustainable decentralized wastewater treatment. *Water*, **16**(20), 3003. https://doi.org/10.3390/w16203003

Shah M. P. (2024). Developments in Wastewater Treatment Research and Processes: Applied Technologies for Clean-up of Environmental Contaminants. Elsevier.

Shah M. P. and Rodriguez-Couto S. (2019). Microbial Wastewater Treatment. Elsevier, Amsterdam, Netherlands.

Silva J. A. (2023). Wastewater treatment and reuse for sustainable water resources management: a systematic literature review. *Sustainability*, **15**(14), 10940. https://doi.org/10.3390/su151410940

Singh P., Bisen M., Kulshreshtha S., Kumar L., Choudhury S. R., Nath M. J., Mandal M., Kumar A. and Patel S. K. S. (2025). Advancement in anaerobic ammonia oxidation technologies for industrial wastewater treatment and resource recovery: a comprehensive review and perspectives. *Bioengineering*, **12**(4), 330. https://doi.org/10.3390/bioengineering12040330

Sravan J. S., Matsakas L. and Sarkar O. (2024). Advances in biological wastewater treatment processes: focus on low-carbon energy and resource recovery in biorefinery context. *Bioengineering*, **11**, 281. https://doi.org/10.3390/bioengineering11030281

Subeesh A. and Chauhan N. (2026). Agricultural digital twin for smart farming: a review. *Green Technologies and Sustainability*, **4**(2), 100299. https://doi.org/10.1016/j.grets.2025.100299

Suprun E., Mostafa S., Stewart R. A., Villamor H., Sturm K. and Mijares A. (2022). Digitisation of existing water facilities: a framework for realising the value of scan-to-BIM. *Sustainability*, **14**(10), 6142. https://doi.org/10.3390/su14106142

Szpyrkowicz L., Kaul S. and Neti N. (2005). Tannery wastewater treatment by electro-oxidation coupled with a biological process. *Journal of Applied Electrochemistry*, **35**, 381–390. https://doi.org/10.1007/s10800-005-0796-7

Tamburus I., Oliveira C., Rodrigues W. and Anhê A. (2025). Efficiency of both activated sludge system and its variations in removing the chemical oxygen demand from industrial waste: systematic review and meta-analysis. *Ciência E Natura*, **47**, e84287. https://doi.org/10.5902/2179460X84287

Tao F., Zhang H., Liu A. and Nee A. Y. C. (2019). Digital twin in industry: state-of-the-art. *IEEE Transactions on Industrial Informatics*, **15**(4), 2405–2415. https://doi.org/10.1109/TII.2018.2873186

Tee F., Abdullah M., Wei Tan I. A., Khairunnisa N. F., Afizal M., Nolasco-Hipolito C. and Bujang K. (2016). Review on hybrid energy systems for wastewater treatment and bio-energy production. *Renewable and Sustainable Energy Reviews*, **54**, 235–246. https://doi.org/10.1016/j.rser.2015.10.011

Toth A. J., Fozer D., Mizsey P., Varbanov P. and Klemeš J. (2022). Physicochemical methods for process wastewater treatment: powerful tools for circular economy in the chemical industry. *Reviews in Chemical Engineering*, **39**, 1–29. https://doi.org/10.1515/revce-2021-0094

Viswanathan S. and Manisankar P. (2025). Electrochemical techniques in sustainable water management. In: S. Patra, J. G. Manjunatha and M. Sillanpää (eds), Climate-Neutral Water Management: A guide to sustainable solutions, IOP Publishing, Bristol, UK, pp. 5-1–5-30. https://doi.org/10.1088/978-0-7503-6382

Vlachou V. I., Karakatsanis T. S. and Efstathiou D. E. (2025). Recent advances of artificial intelligence methods in PMSM condition monitoring and fault diagnosis in elevator systems. *Applied System Innovation*, **8**(5), 154. https://doi.org/10.3390/asi8050154

Wang A.-J., Li H., He Z., Tao Y., Wang H., Yang M., Savic D., Daigger G. T. and Ren N. (2024). Digital twins for wastewater treatment: a technical review. *Engineering*, **36**, 21–35. https://doi.org/10.1016/j.eng.2024.04.012

Worku A. K., Ayele D. W., Teshager M. A., Omar M., Yerkrang P. P., Elgaddafi R. and Alemu M. A. (2025). Recent advances in wastewater treatment technologies: innovations and new insights. *Energy Reviews*, **4**(4), 100164. https://doi.org/10.1016/j.enrev.2025.100164

Yang S., Behzadian K., Coleman C., Holloway T. G. and Campos L. C. (2025). Application of AI-based techniques for anomaly management in wastewater treatment plants: a review. *Journal of Environmental Management*, **392**, 126886. https://doi.org/10.1016/j.jenvman.2025.126886

Yeasmin F., Rasheduzzaman M., Manik M. and Mehedi M. (2023). Activated sludge process for wastewater treatment. In: Advanced and Innovative Approaches of Environmental Biotechnology in Industrial Wastewater Treatment, Springer Nature Singapore, pp. 23–50. https://doi.org/10.1007/978-981-99-2598-8_2

Yusuf F. G., Rasheed A. A., Ibrahim A. B. and Ogunleye S. (2026). Energy and resource recovery from tannery wastewater for sustainable leather production. *Discover Environment*, **4**(1), 39. https://doi.org/10.1007/s44274-026-00517-6

Zhiji D., Caliendo L., Panico A., Esposito G., van Hullebusch E. and Pirozzi F. (2015). An innovative approach to remove nitrogen from wastewater using a biological ANaerobic AMMonium OXidation (ANAMMOX) process. *Chemical Engineering Transactions*, **43**, 2245–2250. https://doi.org/10.3303/CET1543375

Zamfir F.-S., Carbureanu M. and Mihalache S. F. (2025). Application of machine learning models in optimizing wastewater treatment processes: a review. *Applied Sciences*, **15**(15), 8360. https://doi.org/10.3390/app15158360

Zhan J., Yao Y. and Wang X. (2025). Electrochemical technologies for sustainable agricultural water treatment and resource recovery. *International Journal of Electrochemical Science*, **20**(7), 101029. https://doi.org/10.1016/j.ijoes.2025.101029

Zhang P., Shen D., Wang X., Wu S., Long Y. and Gu F. (2025). Enhanced anaerobic digestion of waste activated sludge using magnetite-modified sludge ceramsite: performance and microbial dynamics. *Journal of Environmental Management*, **376**, 124389. https://doi.org/10.1016/j.jenvman.2025.124389

Zhang D., Liu J., Wang H. and Wang Y. (2026). Advancing carbon-neutral wastewater treatment: artificial intelligence-driven strategies for emission mitigation and process optimization. *Environmental Research*, **290**, 123449. https://doi.org/10.1016/j.envres.2025.123449

Zheng T.-H., Zhang Z.-Z., Liu Y. and Zou L.-H. (2025). Recent progress in catalytically driven advanced oxidation processes for wastewater treatment. *Catalysts*, **15**(8), 761. https://doi.org/10.3390/catal15080761

Zieliński M., Kazimierowicz J. and Dębowski M. (2023). Advantages and limitations of anaerobic wastewater treatment–technological basics, development directions, and technological innovations. *Energies*, **16**(1), 83. https://doi.org/10.3390/en16010083

Zou S., Ju H. and Zhang J. (2025). Water quality management in the age of AI: applications, challenges, and prospects. *Water*, **17**, 1641. https://doi.org/10.3390/w17111641

Zulhendri F., Stuckler D., Simon A., Arief J., Laupe A., Zulhendri F. and Pagilla K. (2025). Long-term performance and fouling mitigation in industrial MBR systems: A systematic review of strategies and trade-offs. *Desalination and Water Treatment*, **324**, 101555. https://doi.org/10.1016/j.dwt.2025.101555

doi: 10.2166/9781789065404_0063

Chapter 4

Phytoremediation of wastewater via aquatic macrophyte biomass and its valorization into nutrient-enriched biofertilizers and biofuels

Sk Md Asif[1], Arup Sen[2], Argha Mandal[3] and Akbar Hossain[4]*

[1]Department of Agriculture, Brainware University, Barasat, Kolkata, India, 700125
[2]Department of Agricultural Chemistry and Soil Science, Bidhan Chandra Krishi Viswavidyalaya, Haringhata, Nadia, India, 741252
[3]Department of Soil Science and Agricultural Chemistry, Visva-Bharati University, Bolpur, India, 731235
[4]Soil Science Division, Bangladesh Wheat and Maize Research Institute, Dinajpur 5200, Bangladesh
**Corresponding author: skmdasifbckv@gmail.com*

ABSTRACT

Freshwater shortage impacts more than one billion people worldwide, while eutrophication persists to degrade aquatic ecosystems by encouraging excess growth of aquatic macrophyte biomass (AMB). Phytoremediation using free-floating plants like water hyacinth (*Eichhornia crassipes*), duckweed (*Lemna* spp.), and water lettuce (*Pistia stratiotes*) has emerged as a viable, cost-efficient technique for removing nutrients, heavy metals, and organic contaminants from wastewater. The species are characterized by rapid growth and high biomass yield, making them suitable for large-scale remediation. However, the post-treatment accumulation of nutrient-rich APB causes significant disposal problems. The chapter explores a circular bioeconomic model that utilizes harvested aquatic biomass to produce value-added products like biochar, nutrient-rich compost, and biofuels. Nitrogen-rich keratinous waste (e.g., hair, wool) and phosphorus-rich bone meal can also be added to composting to raise the agronomic value. Simultaneously, bioenergy can be derived through anaerobic digestion, briquetting, or pyrolysis. Due to its lignocellulosic structure and high carbohydrate content, the biomass is also suitable for producing animal and fish feed, ethanol, and fiberboards. The chapter provides a detailed look at how plants can clean up the environment, ways to make the most of biomass, and the future of treating wastewater while recovering resources. It emphasizes techno-economic feasibility, environmental benefits, and the practicality of macrophyte-based products. Technical and policy limitations currently constrain large-scale implementation, despite its promising potential. Using invasive aquatic plants in these systems helps clean wastewater and provides sustainable materials, making it important for managing water resources and supporting a circular economy.

Keywords: Wastewater treatment, aquatic macrophyte biomass, phytoremediation, nutrient recovery, compost, biofuels

4.1 INTRODUCTION

The global increase in wastewater contamination due to excessive nutrients, especially nitrogen (N) and phosphorus (P); toxic heavy metals, such as cadmium (Cd) and lead (Pb); and persistent organic pollutants, is primarily due to rapid urbanization, population growth, and intense agricultural and industrial activities. These factors deteriorate the quality of water and soil and pose risks to human and animal health. While conventional wastewater treatment processes such as activated sludge processes, membrane filtration, and chemical precipitation are commonplace, they may prove costly due to high operating expenses, involve complex infrastructure, and result in secondary pollution issues such as sludge accumulation and improper contaminant removal (Casini *et al.*, 2021; Cui *et al.*, 2024; Mikajlo *et al.*, 2024).

In this context, phytoremediation has emerged as a cost-effective, energy-efficient, sun-powered, and nature-friendly way to remediate wastewater with very little environmental impact and several economic advantages. Recent studies have shown that aquatic macrophytes are capable of remediating water pollution via non-sustained processes. This situation establishes a research direction for aquatic macrophytes and has vested in phytoremediation. Since the end of the twentieth century, researchers have investigated aquatic macrophytes and their ability to remove nutrients, heavy metals, and organic pollutants from wastewater and polluted aquatic systems (Alexandre *et al.*, 2025; Thampatti *et al.*, 2020). For example, species such as *Eichhornia crassipes*, *Lemna minor*, *Pistia stratiotes*, and *Salvinia molesta* are starting to become the focus of phytoremediation research as they reproduce rapidly and have high biomass yields. In several studies, aquatic macrophytes have been proven to reduce biochemical oxygen demand (BOD), N, P, and certain heavy metals (Mustafa & Hayder, 2020; Wibowo *et al.*, 2023). In addition, practical applications involving the use of aquatic macrophytes are either cost effective or environmentally sustainable options that can be used as substitutes for conventional wastewater treatments (Coimbra & Borges, 2023; Padme, 2024).

However, adapting aquatic phytotech to general phytoremediation applications faces the challenge of post treatment, especially the disposal of large biomass yields that contain several important contaminants. Inappropriate disposal of such contaminated biomass can reintroduce pollutants into the environment, thus defeating the purpose of phytoremediation. This offers opportunities for valorization pathways that transform aquatic macrophyte biomass (AMB) into useful biobased products to complete the nutrient cycle with a lesser effect on the ecosystem (Alhajeri *et al.*, 2024; Felicien *et al.*, 2024; Feng *et al.*, 2024).

Current studies indicate that combined composting and vermicomposting systems that blend aquatic macrophytes with keratinous wastes (i.e., feathers, wool, and hair) and bone meal can produce nutrient-rich compost with high agronomic potential (Reddy *et al.*, 2021; Waliczek *et al.*, 2023).

These waste-derived products serve as effective sources of N and P, improving compost maturity and humification and increasing microbiological activity (Barrena *et al.*, 2007; Rajput *et al.*, 2024). Likewise, biochar produced via the pyrolysis or torrefaction of AMB yields superior compost, nutrient retention, and carbon storage (İlay, 2024; Khan *et al.*, 2023). Concurrently, anaerobic digestion, briquetting, and hydrothermal carbonization (HTC) comprise a collection of bioenergy recovery technologies to convert AMB into a range of renewable energy products, such as biogas and briquetted solid fuel and hydrochar. Such products offer a decentralized energy supply and reduce greenhouse gas emissions (Elsalamony *et al.*, 2024; Kabir, 2015; Li *et al.*, 2024). Additionally, these technologies can yield subsequent coproducts, such as digestate that can be utilized as a fertilizer, and help to achieve a circular bioeconomic model (Ingle *et al.*, 2024).

Although some progress has been made, many of the gaps in the synergistic use of aquatic biomass, keratinous waste, and bone meal still persist. In addition, various barriers need to be addressed while retaining nutrients, managing the quality of biochar, and improving bioenergy systems (Jamil *et al.*, 2024; Mikajlo *et al.*, 2024; Stegenta-Dąbrowska *et al.*, 2024). Additionally, little data are available on the relative performance of a few bioenergy processes based on their energy yields, hence restricting their scalability and policy development (Kaushik *et al.*, 2024; Yin *et al.*, 2020).

This chapter explores the phytoremediation potential of aquatic macrophytes for treating wastewater and addresses the possible applications of the resulting biomass. This research focuses on composting and vermicomposting of keratinous wastes and bone meal, biochar and hydrochar production, anaerobic digestion for biogas production, and other uses such as feed, biostimulants, and adsorbents. This chapter presents an integrated framework for circular bioeconomy that links wastewater treatment, biomass valorization, and agricultural reuse. This chapter presents a compilation of current research findings and case studies aimed at developing a holistic pathway to create robust, economically viable, and environmentally sustainable approaches to wastewater management and resource recovery on the basis of existing aquatic plant biomass.

4.2 AQUATIC MACROPHYTES IN PHYTOREMEDIATION

Aquatic macrophytes are highly effective in phytoremediation because of their ability to adsorb, accumulate, and detoxify a wide range of contaminants such as nutrients (N and P), metals (Cd, Pb, Zn and Cr), and several organic contaminants from treated wastewater. The fast growth rate, high biomass production, and morphological ability of these bacteria to adapt to nutrient-rich or nutrient-polluted environments make them especially useful for wastewater treatment systems in a variety of environments.

4.2.1 Classification of hydrophytes, or aquatic macrophytes

Hydrophytes, or aquatic macrophytes, which belong to a group widely studied by ecologists, are often the greatest contributors to aquatic ecosystems. Macrophytes can be classified in several different ways, but the most commonly used classification is based on the way they grow and where they are located. In this way, macrophytes can be classified into four main classifications: emergent, floating, submerged, and free-floating (Figure 4.1).

4.2.1.1 Emergent macrophytes

Emergent macrophytes have their roots embedded in the substrate, but their leaves and stems are above the water surface and are usually found in the shallow parts of lakes and wetlands. These plants are important because they provide homes and food to many different kinds of animals that live in water and on land. They also help keep sediments in place and stop erosion (Jian, 2018; Siddiqui *et al.*, 2023). Plants in the *Cyperaceae* family are common in various bodies of water (Deshmukh *et al.*, 2016).

4.2.1.2 Floating macrophytes

Floating macrophytes can be further classified into rooted, and free-floating macrophytes. Rooted or anchored macrophytes are plants, such as water lilies, whose roots are stuck to the substrate, whereas their leaves float on the surface of the water. These plants are the most important in aquatic ecosystems because they provide shade, lower water temperature, and prevent phytoplankton blooms by blocking sunlight (Jian, 2018; Rejmankova, 2011). Some examples include water lilies and other floating-leaved plants (Harney, 2014).

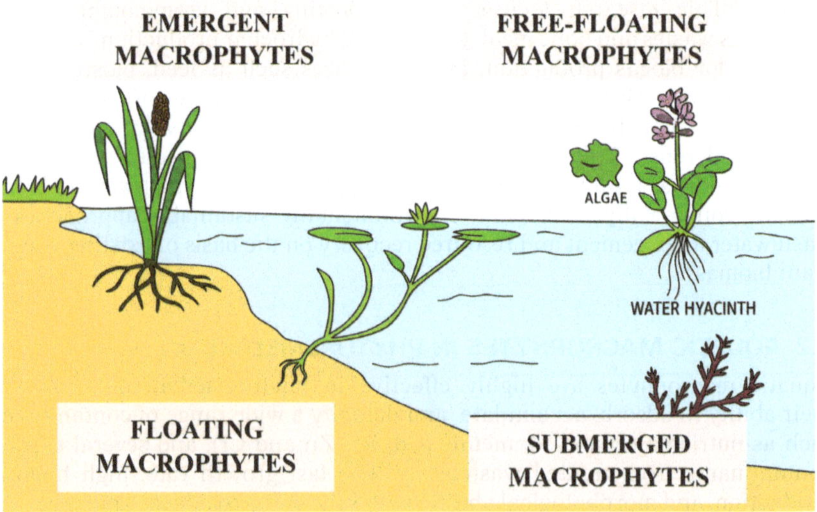

Figure 4.1 Classification of aquatic macrophytes based on growth form and habitat.

4.2.1.3 Submerged macrophytes

Submerged macrophytes are plants that are always completely submerged underwater, and only blooms or reproductive parts will sometimes appear above the surface. These plants play important roles in adding oxygen to the water, providing habitats for living aquatic organisms, and indicating the purity of water (Altinsaçli *et al.*, 2014; Siddiqui *et al.*, 2023). One of the most recognized examples of this species is *Ceratophyllum demersum* (Altinsaçli *et al.*, 2014).

4.2.1.4 Free-floating macrophytes

Free-floating macrophytes are plants that are not attached to the bottom and float freely on the surface of the water. This kind of macrophyte can form large expanses over the water body, which can hinder light penetration into the water, gas exchange, and greater volume of water that aquatic plants occupy. These plants effectively accumulate nutrients and can be used to purify contaminated water (Dijoo *et al.*, 2020). Water hyacinth and duckweed are two fast-growing plants that can potentially absorb nutrients (Dijoo *et al.*, 2020).

These plants that inhabit water are able to adapt in numerous ways to thrive in various forms of water. Environmental variables such as light, temperature, and nutrient availability can directly affect the spatial distribution of macrophytes and the ecological functions they perform (Dar *et al.*, 2014). For example, macrophytes can alter the assemblage in terms of both number and type (Pandey, 2022). Understanding the diversity of macrophytes and their effects on the physical environment is the first step in conserving aquatic ecosystems.

4.2.2 Mechanisms of phytoremediation by aquatic macrophytes in contaminated water bodies

Aquatic macrophytes are important for phytoremediation, a type of green technology that uses plants to remove, stabilize, or degrade environmental pollutants. Aquatic macrophytes are capable of using phytoremediation mechanisms (or processes) such as phytoextraction, rhizofiltration, phytodegradation, and phytostabilization to remediate contaminated water bodies (Figure 4.2).

4.2.2.1 Phytoextraction

Phytoextraction may be defined as the accumulation and uptake of pollutants by the plant's biomass, mostly in the aerial biomass of the plant and mostly heavy metals such as As, Cd, Pb, and Zn (Sinha *et al.*, 2021). Water macrophytes with high enough bioaccumulative potential work well for bioextraction of heavy metals (Mohanty, 2023; Roy *et al.*, 2021).

4.2.2.2 Rhizofiltration

Rhizofiltration is the process of using the root system of a plant to adsorb or precipitate out pollutants onto the plant. This technology can be used for biosorption or precipitation of heavy metals in water. Free-floating macrophytes such as *Eichhornia crassipes* and *Lemna minor* are widely utilized with rhizofiltration because they possess significant root systems which typically lead to improved capability for contaminant adsorption (Mohanty, 2023).

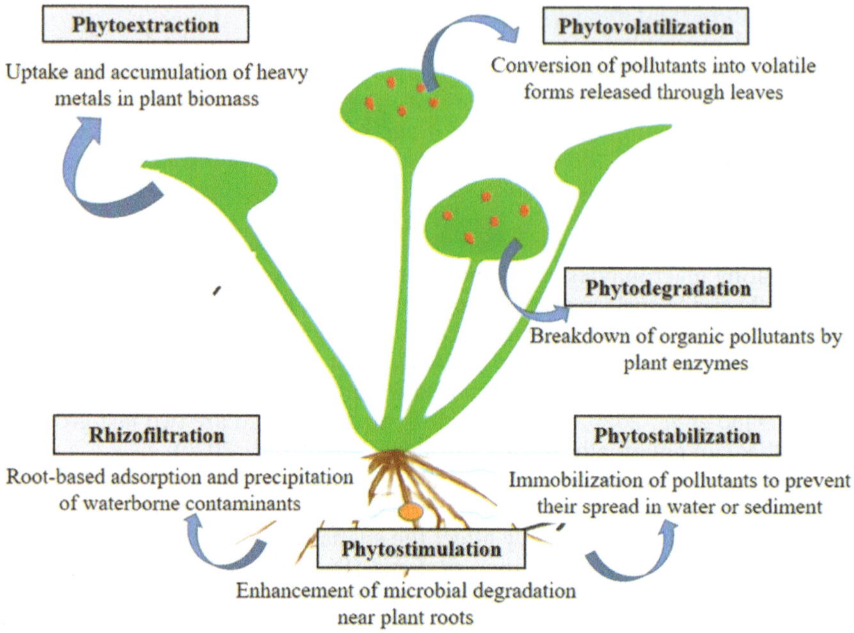

Figure 4.2 Mechanisms of phytoremediation by aquatic macrophytes in contaminated water bodies.

4.2.2.3 Phytodegradation

Phytodegradation (or, phytotransformation) is another mechanism of degrading organic pollutants (often with plant enzymes) of which heavy metals are not commonly associated with such a process, but it is an important mechanism of degrading organic contaminants (e.g., pesticides). There are limited examples of the use of aquatic macrophytes for phytodegradation of metals, however, the general ability of these plants to transform organic pollutants has implications for broader phytoremediation practices (Anand *et al.*, 2023; Pang *et al.*, 2023).

4.2.2.4 Phytostabilization

Phytostabilization involves the immobilization of contaminants already introduced into soil or water to prevent their migration. This mechanism is important for stabilizing metal(loid)s at contaminant sites. Emergent macrophytes such as *Typha angustifolia* and *Phragmites australis*, have been shown to stabilize metals such as As, Cd, and Pb in their root zones and therefore reduce the bioavailability of these elements and their impact on the larger environment (Thampatti *et al.*, 2020; Velasco-Arroyo *et al.*, 2024).

Aquatic phytoremediation is characterized mainly by the four abovementioned processes, but additional processes such as phytovolatilization and phytostimulation may act in some situations. Some macrophytes, for example, have been found to volatilize trace metals such as selenium, and root exudates can also help degrade organic substances through microbial action at sediment-root interfaces. Such processes are generally less common in aquatic environments, and continue to be researched. Thus, aquatic macrophytes can provide a sustainable and high-value solution for remediating water bodies polluted with contaminants. The only downside to using aquatic macrophytes is the disposal of contaminated biomass, which must be managed properly to avoid secondary pollution (Kumar *et al.*, 2022).

4.2.3 Case studies or experiments with pollutant removal efficiency

Aquatic macrophytes have shown extensive potential in the phytoremediation of wastewater because of their ability to remove organic pollutants, heavy metals, and nutrients. These plants operate through different mechanisms including sequestration, uptake, degradation, and rhizofiltration depending on the plant species and type of contaminants present in the water. Current empirical research looks at the performance of floating macrophytes (e.g., *Lemna gibba*, *Pistia stratiotes*), emergent plants (e.g., *Typha latifolia*), and composite systems including native and engineered plants. The removal efficiencies are good, generally in excess of 80% under the best conditions, although the results can be affected by environmental conditions such as pH, temperature, load of pollution, and plant decay. The pollutant removal efficiencies for various aquatic macrophytes are summarized in Table 4.1.

Table 4.1. Pollutant removal efficiencies of various aquatic macrophytes.

Plant Species/ System	Pollutants Targeted	Removal Efficiency	References
Pistia stratiotes L., *Lemna gibba* L.	Heavy metals, phenolic compounds	83–88%	Nafea (2019)
Duckweed (*Lemnaceae*)	Heavy metals (e.g., Pb, Cd, Zn)	Effective accumulation from lake and return waters	Töre and Özkoç (2022)
Canna indica, *Typha latifolia* (vegetated ditch)	Herbicides like Atrazine, Diuron	Atrazine: 43–73%; Diuron: 53–56%	Ling *et al.* (2025)
Water lettuce (*Pistia*), Duckweed (*Lemna*)	BOD, COD	BOD: 88–89%; COD: 66–77%	Mahajan *et al.* (2023)
Pistia stratiotes, *Salvinia* sp.	Total N (TN), Total P (TP)	TN: 86.47%; TP: 75.60%	Cho *et al.* (2013)
Ipomoea aquatica + Bionic macrophyte system	TN, NH_4^+–N, TP, COD	Enhanced over monocultures	Bedabati Chanu and Gupta (2016)
Native macrophyte assemblage	TN, TP, COD	High but marginal over controls; decay issues noted	Song *et al.* (2023)

Together, they identify the promise of phytoremediation with aquatic macrophytes as a low-cost, solar-powered technology for a range of water pollutants. However, optimized outcomes depend on proper species choice, vigilant control of growth conditions, and frequent harvesting to prevent biomass degradation and secondary pollution. Continuing field experimentation and hybrid system design will be needed to increase the consistency and scalability of these nature-based solutions.

4.2.4 Factors affecting performance of aquatic macrophytes

The success of aquatic macrophyte phytoremediation depends on various factors including habitat, plants, and contaminants. To properly construct wastewater treatment systems using macrophytes it is important to know all of the above.

4.2.4.1 Environmental conditions

Factors such as pH, temperature, dissolved oxygen, and nutrient level strongly influence the accumulation of pollutants. High levels of many metals, including Fe and Cu, can inhibit the bioconcentration of some species, whereas pH and oxygen levels can positively affect metal mobility and uptake (Coimbra & Borges, 2023). The BOD, electrical conductivity, N and P load are examples of water quality indicators, and they can reflect how well plants such as *Eichhornia crassipes* and *Pistia stratiotes* can metabolize water (Alexandre *et al.*, 2025).

4.2.4.2 Species-specific characteristics

Different macrophytes present different bioconcentration factors (BCFs) and translocation factors (TFs) and vary in their ability to accumulate and mobilize metals. *Schoenoplectus articulatus* demonstrates superior metal uptake capabilities, whereas *Colocasia esculenta* serves as a more effective stabilizing plant (Khatun *et al.*, 2016; Rasool *et al.*, 2023). Compared with floating plants, emergent plants exhibit a greater capacity for the removal of metals, including Fe, Cu, and Mn.

4.2.4.3 Type and amount of contaminants

The type and quantity of pollutants play a significant role in influencing the efficiency of plants. Some natural macrophytes (Rasool *et al.*, 2023; Saidin *et al.*, 2014) can effectively remove macronutrients such as sulfate and potassium, whereas others such as *Eichhornia crassipes* can efficiently accumulate heavy metals such as Cd, Pb, and Zn. However, high concentrations of pollutants can decelerate the metabolism of plants and reduce their efficiency. Although these traits appear promising, there are nonetheless issues such as interference from numerous contaminants, plant decay, and scalability of the system.

To maximize the utilization of phytoremediation in sustainable wastewater treatment, it is crucial to address such concerns by using appropriate plants, controlling the environment, and incorporating phytoremediation with constructed systems.

4.3 POST-REMEDIATION CHALLENGES

Aquatic macrophyte-based phytoremediation is a high-performance, inexpensive method for the removal of excessive nutrients, heavy metals, and organic pollutants from wastewater. The major post-remediation problem is the accumulation of the AMB, which, if not handled well, can serve as a secondary source of environmental and public health problems.

Additionally, biomass that is not harvested will decay anaerobically which will result in methane release and mobilization of sequestered environmental contaminants, primarily toxic metals. High biomass growth generally results in the development of conditions for standing water, which permits mosquito breeding causing further health issues due to the increased risk of vector-borne diseases. Biomass also interferes with system efficiency by hindering the flow of water due to pathogens in wetlands or drain regions. This ambiguous performance promotes constant upkeep, induces constant operating expenses, and may result in a reinfiltration of pollutants into the surrounding environment with severe regulatory consequences.

The potential consequences of the current post-remediation outcomes are outlined in Table 4.2 and prominent consequences and issues are uncontrolled risks related to AMB. For this reason, it must be treated as a commodity/resource and not waste to help to manage these consequences. In a circular bioeconomy, valorization after decanting by composting, biochar, and conversion to bioenergy is environmentally sustainable and commercially viable.

Table 4.2. Important post-remediation challenges and implications of AMB accumulation.

Challenge	Specific Issues	Implications	References
Environmental risks	Anaerobic decomposition of biomass	Methane emissions and their effects on the climate	Tasnuva (2025)
	Decomposition of metal-rich biomass	Re-release of heavy metals	Patra *et al.*, (2025), Song& Park (2017)
Public health risks	Biomass-induced stagnant water	Breeding mosquitoes; malaria, dengue risk	Patra *et al.* (2024)
Operational burdens	Biomass clogging treatment systems	Blocking flow; lower efficiency; high maintenance	Sharma *et al.* (2022)
Economic consequences	Disposal costs; potential regulatory violations	Increased costs; environmental liability	Song and Park (2017)
Need for valorization	Lack of sustainable biomass use	Lost resource recovery opportunities	Edgar *et al.* (2021), Patra *et al.* (2024)

4.4 VALORIZATION PATHWAYS OF AQUATIC MACROPHYTE BIOMASS

AMB accumulation after phytoremediation is a two-edged sword: it poses the disposal problem but also allows for sustainable recovery of resources. The valorization of AMB is the process of converting this nutrient but potentially toxic biomass into useful products such as biofertilizers, biochar, and biofuels, thereby linking waste management with a circular bioeconomy. Different technological pathways have also been studied and evolved to optimize the nutrient recovery, energy value, and soil fertility enhancement potential of AMB products (Figure 4.3).

4.4.1 Composting and vermicomposting for nutrient-enriched biofertilizer production

Co-amending AMB composting with keratinous wastes such as hair, wool, and feathers, and bone meal has been demonstrated to have great potential in the production of nutrient-rich and stable biofertilizers. Keratinous waste is a source of N, whereas bone meal is a P additive, which is advantageous for improving the NPK value and generally enhances the microbial activity of compost and compost quality (Kumar *et al.*, 2021; Nad *et al.*, 2024). Adding biochar to these systems can help store nutrients, reduce leaching, and support the growth of

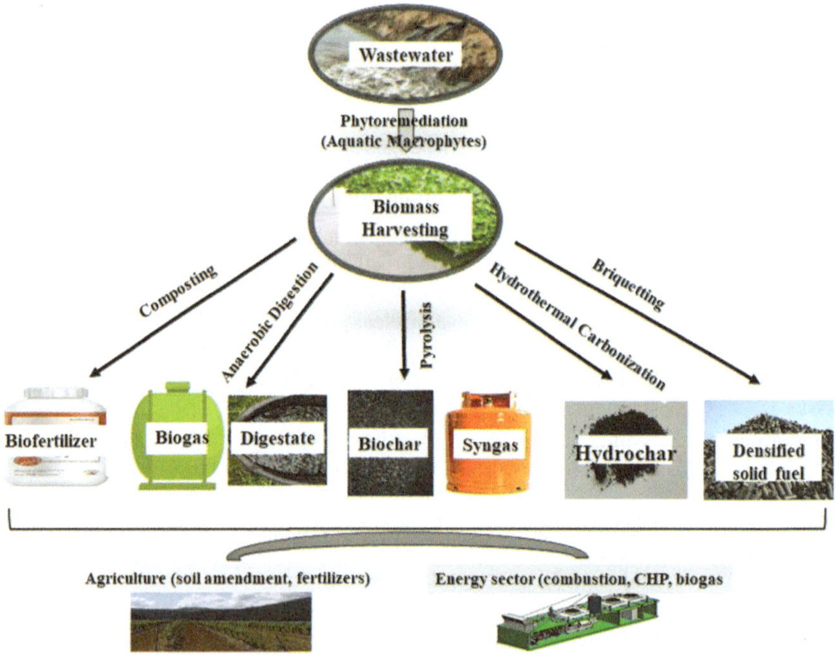

Figure 4.3 Valorization pathways of AMB linking phytoremediation, resource recovery, and circular bioeconomy.

microbial populations and earthworm activities, especially in these systems (Kumar *et al.*, 2022). Several studies reported improved compost maturity parameters, such as maximum C:N ratios (<20), high germination indices (>80), and pathogen suppression by thermophilic temperatures (Waliczek *et al.*, 2023; Yatoo *et al.*, 2022). However, compost quality and nutrient release are determined by the feedstock composition, ratio of amendments, process time, and biochar characteristics, with the predominant influence on P stabilization (Felicien *et al.*, 2024; Feng *et al.*, 2024). Microbial lignocellulose and keratin decomposition by keratinolytic bacteria and fungi, in addition to accelerating humification and nutrient mineralization, support waste detoxification and the development of sustainable biofertilizers, promoting circular bioeconomic tenets (Mehra *et al.*, 2022; Wang *et al.*, 2024).

4.4.2 Biochar production through pyrolysis or torrefaction

The thermochemical conversion of AMB and keratinous wastes by pyrolysis or torrefaction yields nutrient-rich biochar (N, P, and K) with desirable physicochemical characteristics, such as high porosity, surface area, alkaline pH, and cation exchange capacity and thus is a good soil amendment (Cárdenas-Aguiar *et al.*, 2024; İlay, 2024; Khan *et al.*, 2023). Biochar application increases the fertility of soil through increased water-holding capacity, nutrient uptake, microbial biomass, and humification, in addition to promoting long-term carbon sequestration by stabilizing organic carbon in recalcitrant forms (He *et al.*, 2024; Mikajlo *et al.*, 2024). Copyrolysis with additives such as phosphate rock or MgO also increases the retention of nutrients and the immobilization of heavy metals in soils (Rajput *et al.*, 2024). Biochars from aquatic and keratinous biomass have good nutrient contents and structural properties, but differences in feedstock composition and pyrolysis temperature lead to quality and functional variations (Ingle *et al.*, 2024; Kaushik *et al.*, 2024). Recently, machine learning models have been used to rationalize biochar synthesis and model important characteristics in terms of process parameters and improve the accuracy of biochar engineering (Yuan *et al.*, 2024; Zhang *et al.*, 2024a).

There are many short-term benefits, but few long-term field studies on ecological stability, carbon lifespan, and phytotoxicity. Some studies have even reported that pure biochar has negative impacts because harmful products or nutrients are trapped (Mikajlo *et al.*, 2024). Co-composting biochar with organic amendments addresses these risks and increases microbial activity and nutrient cycling, although microbial relationships are complex and underexplored (Ajibade *et al.*, 2022; Casini *et al.*, 2021; Kaushik *et al.*, 2024). In general, biochar from AMB and keratinous residues presents a promising avenue for soil renewal, nutrient cycling, and climate-resilient agriculture in the context of a circular bioeconomy.

4.4.3 Anaerobic digestion for biogas production

Anaerobic digestion of AMB, particularly co-digestion with keratinous wastes such as feathers, is an inexpensive option for the recovery of renewable energy, and up to 0.67 N m^3/kg of methane has been reported (Elsalamony *et al.*, 2024; Forgács *et al.*, 2011; Kabir, 2015). Nonetheless, microbial degradation of certain aquatic and keratin-based feedstocks is challenging because of their lignocellulosic materials and recalcitrant components. Some pretreatment methods–thermal, chemical, enzymatic, and hydrothermal hydrolysis–have been introduced to increase substrate solubilization, to improve the C/N ratio and microorganism accessibility for COD removal (up to 93%) and biogas stability (Alhajeri *et al.*, 2024; Li *et al.*, 2024). These processes lead to the production of nutrient-rich digestates along with hydrogen and methane. The stable digestate is a class of biofertilizer; it also supports nutrient recycling and contributes to a circular bioeconomy.

4.4.4 Briquetting and pelletization for solid biofuel

The briquetting and pelletization of dry AMB is a decentralized low-cost process in which the biomass of moderate-to-high caloric values is changed into solid fuels. Compared to dried biomass, compound biofuels with carbonaceous co-substrates are easier to handle, transport, and store and can be used for rural energy use, such as cooking and small-scale industrial heaters (Amit *et al.*, 2024; Bolognesi *et al.*, 2021). Water hyacinth, duckweed and Salvinia are known to have been depleted. The amount of energy returned varies depending on the feedstock type and how well it is dewatered and compacted. This use is especially suitable for areas with no access to liquid or gaseous fuels, helping people become more energy independent and reducing the amount of biomass waste.

4.4.5 Hydrothermal carbonization for hydrochar and soil amendment

HTC is a promising valorization method for wet AMB, eliminating the energy-intensive drying operations required in pyrolysis or briquetting. At subcritical temperatures (180–250°C), HTC converts biomass into hydrochar–a carbonaceous, energy-intensive material with good fuel quality and the potential to act as a soil amendment (Alhajeri *et al.*, 2024). Hydrochar has similar structural advantages over biochar, such as adsorptivity for pollutants and nutrients, and has also been demonstrated to have opportunities to produce biohydrogen when paired with microbial catalysts. In addition, HTC has the potential to lower pathogen load and complement integrated waste treatment systems with a reduced negative impact on the environment. HTC has proven to be credible; however, it has not gained traction enough for field-scale applications and further studies may lead to the optimization of energy use, cost–benefits, and long-term impacts on soils.

4.5 ADDITIONAL APPLICATIONS OF AQUATIC MACROPHYTE BIOMASS

4.5.1 Feedstock for microbial enzyme production

AMB acts as a feedstock to produce microbial enzymes because of its abundance, rapid growth, and similarity to lignocellulosic biomass. Aquatic plants such as Azolla can produce substantial biomass quickly, especially in eutrophic water bodies, without competing for arable land or food resources (Kotamraju *et al.*, 2023; Miranda *et al.*, 2016). These crops have compositions that are akin to those of lignocellulosic biomass but with reduced levels of lignin, making pretreatment, which is involved in enzyme production, easy (Kotamraju *et al.*, 2023; Nargotra *et al.*, 2022). The economic viability of the use of aquatic plants as raw materials relies on the creation of cost-efficient and innovative methods for biomass conversion and enzyme production (Biswal & Mandavgane, 2021; Ravindran & Jaiswal, 2016). In addition, gene modification and the development of strains of microbes propose further improvements in yield and the efficiency of the enzyme production of aquatic biomass to become a sustainable option in the bioeconomy.

4.5.2 Bioplastics and biofilms

Bioplastics and biofilms are considered both economically and ecologically when the AMB content is used. The abundance of cellulose and fast growth make macroalgae good bioplastic sources (Elkaliny *et al.*, 2024). The use of aquatic biomass to produce nonpetroleum-derived bioplastics is economically viable, as renewable feedstock is utilized and cost savings are obtained through the utilization of waste biomass. In inventive ways, their industrial wastes could be used as carbon sources in production (soil conditioner molasses, palm oil waste effluent, etc.) thus avoiding treatment and disposal costs (Ibrahim *et al.*, 2017). They are also compatible with existing infrastructure because their properties are similar to those of petroleum plastics, increasing their market value (Rahman & Miller, 2017). Bioplastics are also eco-friendly because they release lower amounts of emissions and are biodegradable, thus minimizing environmental degradation resulting from the use of regular plastics. Despite this, however, scalability issues, economic feasibility, and technological impediments (Narayanasamy *et al.*, 2024; Ugya *et al.*, 2023), there could be support from research and development, transition policies, and market support to enable the transition to a greener bioplastics sector. The applicability of a circular economy models can enhance their contribution to the sustainability of the environment.

4.5.3 Heavy metal removal adsorbents

The potential use of AMB, or specifically, dead biomass consisting of algae, seaweed and watermilfoil, offers an effective and cost-effective alternative to conventional adsorbents for removing heavy metals from wastewater. Given their high growth rates, extensive distribution, and tendency to be invasive, aquatic plants have a twofold environmental advantage—allowing for pollutant uptake while controlling ecological disturbances (Singh & Kumar, 2022; Znad *et al.*, 2022). Ion-exchange, surface complexation, and precipitation-induced biosorption occur via surface functional groups such as carboxyl, hydroxyl, and amino groups on plant biomass surfaces that chelate heavy metal ions effectively (Karnwal *et al.*, 2024). Additionally, AMB treatment based on chemicals or nanotechnology greatly increases the surface area and adsorption capacity (Zhang *et al.*, 2024b). Marine macroalgae such as *Desmodesmus* sp. possess excellent metal biosorption capacity for lead and cadmium with reusability established through multiple cycles (Waqar *et al.*, 2023; Znad *et al.*, 2022). In addition, agricultural waste, such as modified corn husk fibers has also been shown to be promising, widening the scope of biomass-based adsorbents beyond water sources (Zhang *et al.*, 2024b). However, biosorption systems are pH-sensitive, metal concentration-sensitive, and contact time-sensitive and require optimal process adjustment (Jain *et al.*, 2016). While exploiting invasive species is advantageous, it must be controlled to avoid unintended consequences to the ecosystem. In a broad context, AMB biostabilization of biosorbents adheres to sustainable wastewater treatment goals, through environmentally superior, efficient, and broad-scale remediation of heavy metals.

4.5.4 Biopesticides or biostimulants

Aquatic macrophytes, even microalgae, provides renewable and economically viable feedstock to synthesize bioactive molecules for biostimulants and biopesticides to replace agrochemicals with green alternatives. The biorefinery approach has increased economic feasibility when AMB is used because it is possible to coextract biocrude oil, biofertilizers and plant growth regulators from AMB via the same biomass stream to produce high value products (Mohanty & Mohanty, 2023). Growing organisms such as *Chlorella thermophila* in wastewater not only minimizes the costs of inputs but also helps in the treatment of water, following the tenets of circular economy. Low-cost extraction with green solvents (e.g., ethanol and acetone) has resulted in high biostimulant activity and germination indices in *Scenedesmus almeriensis* (Navarro-López *et al.*, 2023). The annual growth rate of global biopesticide market of 10–15% reflects the demand for less toxic and more sustainable agricultural inputs (Chakraborty *et al.*, 2023). Biopesticides and other pest management products that reduce chemical utilization and risks to people and the environment, can be combined with methods to extract biomass through sustainable sources, and rehydrate and grow in wastewater to reduce resource use and footprints (Mogany *et al.*, 2024). With that said, scaling issues, regulatory challenges, and market penetration are identified concerns. Continued developments in biomass harvesting, formulation, and market/product support policies to be able to utilize and apply it as long-term biopesticides and biostimulants are needed in the future. However, the valorization of AMB products as biopesticides and biostimulants generally occurs in concert with sustainable agriculture and for good reasons for green agro-industrial practices should be considered.

4.5.5 Feed additives

The biomass of aquatic plants such as microalgae and macrophytes such as *Ipomoea aquatica*, *Lemna minor*, *Azolla pinnata*, and *Chlorella* or *Arthrospira* spp. (*Spirulina*), is a very viable option as a green substitute for fishmeal in aquaculture diets. Proteins constitute 25–65% of dry biomass and have a high unsaturated percentage of omega-3 fatty acids, these organisms have also been proven to have comparable or better effects on fish growth, immunity, and feed conversion ratios while using less ecologically unreliable marine-based resources (Omweno *et al.*, 2024; Safafar, 2017). Microalgal culture in aquaculture wastewater yields a doubled biomass yield and nutrient recovery, thus reducing eutrophication and advancing a circular economy (Shaika *et al.*, 2024). Nonetheless, the implications for mass production are the high cost of cultivation, harvesting, and processing as well as competition for freshwater, land, and P when plant-based feed sources are used (Malcorps *et al.*, 2019). Overcoming these constraints will require cooperation among policymakers, farmers, and researchers to improve not only cultivation systems but also alternative feed sources for conventional aquaculture. Despite these constraints, AMB valorization as an aquaculture feed additive is a promising path toward more sustainable and resilient aquaculture.

4.6 INTEGRATED CIRCULAR BIOECONOMY MODEL

Despite the shift to a circular bioeconomy, a new system can likely be developed to incorporate the treatment of wastewater, resource recovery, and renewable energy generation. Aquatic macrophytes and microalgae are at the center of this approach because they are used as phytoremediators, removing contaminants from wastewater, and renewable biomass sources. The hybrid system at work first utilizes nutrient-enriched wastewater–N, P, organic pollutants, and heavy metals–as a growth medium for high-growth aquatic plants. These organisms can remove the pollutants via phytoextraction, rhizofiltration, and biosorption (Lutzu *et al.*, 2024; Shaika *et al.*, 2024).

Instead of discarding polluted biomass after remediation–upon disposal the disposal may create secondary pollution and release potential additional greenhouse gas emissions–we can process the biomass to produce a valuable end product, through a variety of routes. This biomass can be repurposed in a number of ways including composting/ vermicomposting as a biofertilizer (Yatoo *et al.*, 2022), anaerobic digestion for biogas (Forgács *et al.*, 2011), and pyrolysis for biochar (He *et al.*, 2024), hydrothermal carbonation for hydrochar, and briquetting for solid biofuels (Elsalamony *et al.*, 2024). Each pathway provides safe management of the biomass, while providing a net economic and environmental benefit by converting a remediation waste by product into useful agricultural and energy resources.

The products–nutrient-rich composts, renewable energy, and soil conditioners rich in carbon-are brought back into the agriculture and energy sectors, hence closing the loop. This cyclical process not only enhances soil health and diminishes reliance on synthetic fertilizers but also supports climate mitigation by sequestering carbon and utilizing renewable energy (Mohanty & Mohanty, 2023; Rajput *et al.*, 2024). By substituting linear waste management systems, this approach welcomes decentralization, resource circularity, and sustainability.

A conceptual flow diagram is included below to represent the hybrid system in practice (Figure 4.4). This hybrid approach aligns with international initiatives paving the way toward cleaner production, sustainability in agriculture, and development of climate-resilient infrastructure. Nonetheless, the broader adoption of this work will depend on addressing questions of economic viability, biomass conversion technology, and policy for circularity.

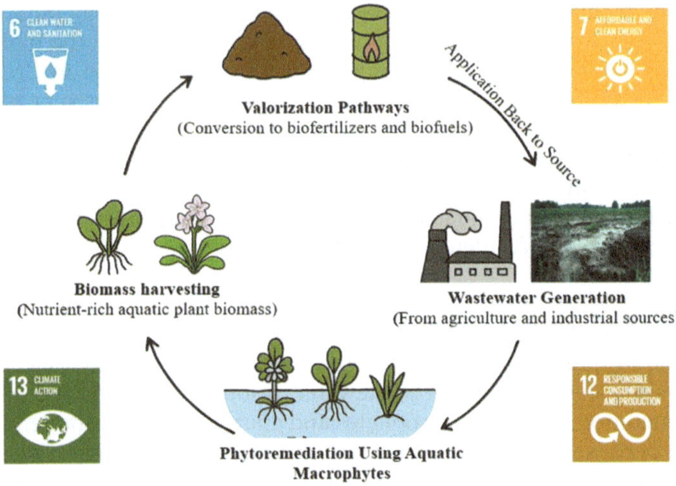

Figure 4.4 Schematic of the integrated circular bioeconomy model for wastewater treatment and biomass valorization.

4.7 COMPARATIVE ASSESSMENT OF VALORIZATION ROUTES: TECHNOECONOMIC AND ENVIRONMENTAL PERSPECTIVES

Compared with fossil fuel alternatives, the sustainability of AMB depends on combinations of valorization routes that are technically feasible, economically viable, scaleable, environmentally virtuous/friendly (biodiversity for society and ecosystem services). With phytoremediation systems generating massive amounts of biomass, the decision on downstream processing options ranging from composting to thermochemical conversion needs to be based on cost-effectiveness, scalability, environmental impact, and end-product characteristics.

Technoeconomic studies confirm that composting, particularly when combined with keratinous or nutrient wastes, is a low-cost, high-impact technology with low technological inputs and is suitable for decentralized rural application. Anaerobic digestion potentially requires moderate capital costs, with two disadvantages: (1) spoilage after a time and (2) the need for control if not cultivated. Furthermore, anaerobic digestion benefits biogas and nutrient-rich digestate. Pyrolysis and HTC for biochar production, benefit longevity and carbon sequestration potential in soil. However, these benefits are associated with increased energy use and emission control. The results of LCA also indicate that while HTC and hydrothermal liquefaction (HTL) offer high carbon recovery and energy efficiency, they may require more freshwater and more chemicals to carry out. In contrast, briquetting is a straightforward valorization method with low emissions but low energy density unless the biomass is pretreated.

Table 4.3 provides an overview of these valorization routes for stakeholders, researchers, and policymakers. This comparison assessment lays the groundwork for an integrated circular bioeconomic model for linking wastewater bioremediation with resource recovery and emphasizes sustainability and practicality.

Table 4.3. Technoeconomic and environmental comparison of AMB valorization pathways.

Valorization Route	Techno-Economic Feasibility	Environmental Performance	Key Outputs	Limitations
Composting/ Co–Composting	High; low setup cost; short payback; viable at community/farm scale (Pandey, 2022; Song, 2020)	Low emissions; promotes nutrient recycling; improves soil fertility	Organic fertilizer (compost)	Limited energy recovery; odor and pathogen control needed
Vermicomposting	Moderate to high; produces enriched compost; efficient at small scale (Alshehrei *et al.*, 2025; Rodríguez-Palacio *et al.*, 2018)	Eco-friendly; enhances microbial diversity and soil health	High-quality biofertilizer	Sensitive to moisture, temperature, and contamination
Anaerobic digestion (AD)	Moderate; revenue from biogas sale offsets costs; digestate adds fertilizer value (Yatoo *et al.*, 2022)	Enables methane recovery; reduces GHG emission; supports nutrient recycling	Biogas, digestate	Requires significant capital investment; digestate treatment needed
Pyrolysis	Moderate; scalable with energy co-recovery systems (Sekar *et al.*, 2021; Sun *et al.*, 2022)	Produces biochar for carbon sequestration; improves soil properties; immobilizes pollutants	Biochar, bio-oil, syngas	High energy demand; strict emission control required
Hydrothermal carbonization (HTC)	Moderate; suitable for wet biomass; enabled by emerging reactor designs (Rajput *et al.*, 2024)	Generates lower GHG emissions than pyrolysis; produces hydrochar for soil and energy uses	Hydrochar, process water	High pressure systems; HTC effluent requires management
Briquetting/ Pelletization	Low to moderate; low-tech; profitable if biomass is pretreated (Charis *et al.*, 2024)	Replaces wood/coal; emits low sulfur oxides (SO_x) and nitrogen oxides (NO_x)	Solid biofuels	Requires drying; may emit particulate matter (PM) if poorly combusted
Hydrothermal liquefaction (HTL)	Low; high capital and operational expenditure (CAPEX/OPEX); not yet commercially viable at scale (Biller *et al.*, 2012)	Converts wet biomass to biocrude oil; offsets fossil fuels but higher eutrophication potential	Biocrude oil, char, aqueous phase	Expensive process; complex purification of end products

4.8 CASE STUDIES AND PRACTICAL IMPLEMENTATIONS

4.8.1 Successful pilot or field-scale projects

Phytoremediation with aquatic macrophytes has exhibited significant potential in the treatment of wastewater in both developing and developed nations. Various pilot- and field-scale applications have resulted in significant improvements in water quality, effective pollutant removal, and ecosystem restoration. A summary of such exemplary case studies on water hyacinth, duckweed, and other aquatic plants is presented in Table 4.4. These case studies illustrate the potential for systems, such as floating wetlands, built wetlands, and integrated fish–plant systems to provide cost-effective, nature-based treatment options, particularly where there are limited infrastructure capabilities.

When moving to field-scale application of macrophytes such as, *Eichhornia crassipes*, *Lemna minor*, and *Pistia stratiotes* the potential for pollutant and nutrient extraction from wastewater can be confirmed, and the economic potential for biomass valorization is highlighted. The structure-rich biomass obtained from harvesting systems containing these types of species can be converted into value added products such as biofertilizers and biofuels. All of these actions provide an integrated approach that supports the circular economy by combining a wastewater treatment option with resource recovery. Overall, phytoremediation provides a low-cost, environmentally friendly, and often easily scalable option for alternative sustainable wastewaters.

4.8.2 Challenges encountered

Despite their advantages, macrophyte-based phytoremediation systems do have some known eco- or practical difficulties and challenges:

- Potentially disturbing species invasion: Although water hyacinth (*E. crassipes*) is an excellent nutrient absorber, it is also an invasive plant. This might be an issue. If left unchecked, it may have an impact on freshwater habitats (Nazir, 2020).
- Land and space limitations: Lack of available land, especially in urban areas of developing countries, makes it less feasible to utilize constructed wetlands in large-scale applications (Tang, 2024).
- Differences in system performance: Phytoremediation is dependent on microbial activity, plant development, and many other factors which also vary at different times during the year. Zheng *et al.* (2016) reported that the HCW system in Xi'an experienced a relatively low rate of contaminant removal at different times of the year.
- Challenges associated with scale-up and public awareness: Many methods that work well on a small scale have never been expanded because there is not enough money, public awareness, or expert availability (Zaman *et al.*, 2024).
- Monitoring and maintenance: Continuous operation involves regularly checking the biomass and frequency of nutrients concentration harvesting, which is more difficult under poorer conditions (Mohan *et al.*, 2016).
- Social acceptance: Social acceptance and the perception of 'plant-based sewage treatment' can discourage adoption. Socioeconomic education

Table 4.4. Selected case studies of aquatic macrophyte-based phytoremediation systems.

Location	Macrophyte/ System	Target Pollutants	Key Outcomes	References
East Kolkata Wetlands, India	Water hyacinth	BOD, COD, heavy metals	Improved water quality; eco-restoration potential	Sanyal *et al.* (2015)
Bhubaneswar, India	Duckweed + fish culture	Ammonia, nitrate, phosphate, BOD	Significant nutrient reduction in pilot-scale integrated system	Jena *et al.* (2011)
Tangxi River, China	Sequential hybrid constructed wetland (SHCW)	COD, NH_4^+-N, TP	Removal efficiencies: COD (72.5%), NH_4^+-N (91.6%), TP (76.3%)	Gan (2022)
Xi'an, China	Hybrid constructed wetland	BOD_5, suspended solids	BOD_5 removal (94.4%), TSS (92.0%); seasonal variation noted	Zheng *et al.* (2016)
Ota, Nigeria	Water hyacinth, duckweed, water lettuce	BOD, COD, coliforms, TSS, *E. coli*	BOD removal (91.3%), COD (85.0%), TSS (96.2%), *E. coli* (94.4%)	Justin and Olukanni (2024)
Niger Delta, Nigeria	Water hyacinth, water lettuce	Chromium, cadmium	Removal efficiencies: Cr (97%), Cd (87.5%)	Bekeowei and Bariweni (2022)
Northern Italy	Macrophyte bed (reeds)	Sludge dewatering, COD, TSS, N	Sludge solids: 23–30%; 90% N removal; compliance with EU Directive 91/271/CEE	Barbieri *et al.* (2003)
Brazil (urban streams)	Water hyacinth, water lettuce	BOD, N, P	Reduced BOD and nutrient loads in tropical streams	Alexandre *et al.* (2025)

campaigns play a significant role in large-scale implementation (Yongabi *et al.*, 2018).

- Suitability of climate and soil: The temperature and pH of the soil are site-specific parameters that influence the effectiveness of macrophyte growth and contaminant absorption (Miranda *et al.*, 2022).

4.9 POLICY AND RESEARCH RECOMMENDATIONS

To increase the successful implementation of aquatic macrophyte phytoremediation technologies, stimulating supportive policies and research interventions is essential. This section presents strategic directions that merge ecological innovation, technological innovation, and the principles of circular economy.

4.9.1 Policy recommendations

- Promote the use of macrophytes in decentralized wastewater treatment systems: Both federal and local governments should encourage the use of decentralized wastewater management systems, such as floating wetlands and built wetlands (CWs), in rural and peri-urban areas. CWs use macrophytes with more biomass, such as *Eichhornia crassipes*, *Lemna minor*, and *Hydrilla verticillata* (Saha *et al.*, 2022), to treat wastewater in an inexpensive and nontechnical way.
- Direct R&D resources toward biomass valorization technologies: Sufficient institutional and financial resources must be mobilized to help develop and valorize the phytoremediated biomass into nutrient-rich biofertilizers and biofuels under a circular bioeconomy (Khan *et al.*, 2023). Policies should dictate the limits of biofertilizers used within crop cultivation systems, to ensure that they are used safely and are of good quality.
- Policy to encourage digital water management: Policies should encourage the use of IoT-based monitoring systems of phytoremediation units to provide real-time data such as pH, temperature, and turbidity data and enhance treatment results while complying with regulations (Noerhayati *et al.*, 2023).

4.9.2 Future research recommendations

- Long-term field trials in various climates and scales: There is an urgent need for statistically rigorous multiseason field studies that evaluate the performance of phytoremediation systems versus hydrological and environmental conditions to support reliable and higher-level scale-up potential trajectories (Thakur *et al.*, 2023).
- Microbial synergy in composting and biochar-enriched AMB: Research should focus on the microbial ecology of composting processes, including AMB composting processes and, particularly, when combined with biochar. This method is capable of improving the recovery of nutrients, immobilizing pollutants, and increasing the agronomic value of the end product (Kumar *et al.*, 2024).
- Integration with IoT and AI for process optimization: Future studies should explore synergistic uses of AI and the IoT for decision automation, system performance forecasting, and optimal biomass harvesting and reuse cycles (Noerhayati *et al.*, 2023).

4.10 CONCLUSION AND FUTURE OUTLOOK

Phytoremediation by aquatic plants is a nature-based, decentralized solution for wastewater treatment, with high sustainability potential in low-resource contexts. Together with biomass valorization efforts–converting harvested biomass into nutrient-rich biofertilizers and biofuels–such highly sustainable systems can significantly contribute to climate change mitigation and rehabilitation while also matching circular economy principles. The embedding of low-cost, multibenefit systems supports international goals for sustainability and can strongly contribute to several United Nations sustainable development goals (SDGs), including Clean Water and Sanitation (SDG 6), Affordable and Clean Energy (SDG 7), Responsible Consumption and Production (SDG 12), and Climate Action (SDG 13). However, for phytoremediation of APB to achieve its fullest potential, it is evident that stronger policies, standardization of biobased products, implementation incentives, and cross-disciplinary research in environmental management, microbiology, agronomy, and engineering must address the same challenges. Phytoremediation utilizing APB has the potential to become a low-cost, environmentally friendly method for pollution mitigation and resource recovery with investment and support from the government.

4.11 AUTHOR CONTRIBUTIONS

Sk Md Asif conceptualized and designed the chapter, undertook the majority of the writing, and compiled the final manuscript; Arup Sen contributed to selected sections, provided methodological inputs, and assisted with literature sourcing; Argha Manda supported the work through paper collection, data compilation, and referencing and Akbar Hossain, as the international co-author, critically reviewed the manuscript and offered expert suggestions. All authors read and approved the final manuscript.

ACKNOWLEDGMENTS

The authors sincerely thank their respective institutions for providing facilities and academic support during the preparation of this chapter.

CONFLICT OF INTEREST

The authors declare no known competing financial interests or personal relationships that could have appeared to influence the work reported in this chapter.

FUNDING

This work did not receive any specific grant from funding agencies in the public, commercial, or not-for-profit sectors.

REFERENCES

Ajibade S., Nnadozie E. C., Iwai C. B., Ghotekar S., Chang S. W., Ravindran B. and Awasthi M. K. (2022). Biochar-based compost: a bibliometric and visualization analysis. *Bioengineered*, **13**(7–12), 15013–15032. https://doi.org/10.1080/21655979.2023.2177369

Alexandre D. S., Ogura A. P., de Almeida Mohedano R. and Thibau L. B. G. (2025). Potential of two floating aquatic macrophytes in improving water quality: a case study in two tropical streams. *Revista Brasileira de Ciências Ambientais*, **60**, e2158. https://doi.org/10.5327/z2176-94782158

Alhajeri N. S., Tawfik A., Al-Fadhli F. M. and Nasr M. (2024). Enhancing hydrogen production and biochar recovery from algal biomass: a novel techno-economic synergism with gelatinous digestate. *Journal of Water Process Engineering*, **61**, 105270. https://doi.org/10.1016/j.jwpe.2024.105270

Alshehrei F., Al-Enazi N. M. and Ameen F. (2025). Vermicomposting amended with microalgal biomass and biochar produce phytopathogen-resistant seedbeds for vegetables. *Biomass Conversion and Biorefinery*, **15**(2), 1795–1802. https://doi.org/10.1007/s13399-021-01770-w

Altınsaçli S., Altınsaçli S. and Paçal F. P. (2014). Species composition and distribution of ostracods (Crustacea, Ostracoda) in some lakes and lagoons (Kocaeli and Sakarya, Turkey). *Journal of Entomology and Zoology Studies*, **2**(5), 182–192.

Amit S., Dharnaik M. and Ahmed A. K. T. (2024). Biochar production of kitchen waste and water hyacinth by low temperature steam torrefaction, pp. 75–89. https://doi.org/10.58532/v3bice4p4ch3

Anand R., Mohan L. and Bharadvaja N. (2023). Phyto- and microbial-remediation of radioactive waste. In: Shah M. P. (ed.), Microbial Technologies in Industrial Wastewater Treatment. Springer, Singapore. https://doi.org/10.1007/978-981-99-2435-6_12

Barbieri A., Garuti G., Avolio F. and Bruni S. (2003). Sludge dewatering using macrophytes in a small wastewater treatment system: a case study of a pilot scale plant in northern Italy. *Journal of Environmental Science and Health Part A-Toxic/Hazardous Substances & Environmental Engineering*, **38**(10), 2425–2433. https://doi.org/10.1081/ESE-120023449

Barrena R., Pagans E., Artola A., Vázquez F. and Sánchez A. (2007). Co-composting of hair waste from the tanning industry with de-inking and municipal wastewater sludges. *Biodegradation*, **18**(3), 257–268. https://doi.org/10.1007/S10532-006-9060-Z

Bedabati Chanu L. and Gupta A. (2016). Phytoremediation of lead using *Ipomoea aquatica* Forsk. in hydroponic solution. *Chemosphere*, **156**, 407–411. https://doi.org/10.1016/j.chemosphere.2016.05.001

Bekeowei R. A. and Bariweni P. A. (2022). A study of aquatic macrophyte for remediation of chromium and cadmium in wastewater effluents in Yenagoa Metropolis, Niger Delta. *Journal of Applied and Natural Science*, **14**(4), 1465–1470. https://doi.org/10.31018/jans.v14i4.3808

Biller P., Ross A. B., Skill S. C., Lea-Langton A., Balasundaram B., Hall C., Riley R. and Llewellyn C. A. (2012). Nutrient recycling of aqueous phase for microalgae cultivation from the hydrothermal liquefaction process. *Algal Research*, **1**(1), 70–76. https://doi.org/10.1016/j.algal.2012.02.002

Biswal D. and Mandavgane S. A. (2021). Biomass waste: a potential feedstock for cellulase production. In: Current Status and Future Scope of Microbial Cellulases, Elsevier, pp. 347–359. https://doi.org/10.1016/B978-0-12-821882-2.00017-X

Bolognesi S., Bolognesi S., Bernardi G., Callegari A., Dondi D. and Capodaglio A. G. (2021). Biochar production from sewage sludge and microalgae mixtures: properties, sustainability and possible role in circular economy. *Biomass Conversion and Biorefinery*, **11**(2), 289–299. https://doi.org/10.1007/S13399-019-00572-5

Cárdenas-Aguiar E., Gascó G., Lado M., Méndez A., Paz-Ferreiro J. and Paz-González A. (2024). Characterization of biochar from beach-cast seaweed and its use for amelioration of acid soils. *Land*, **13**(6), 881. https://doi.org/10.3390/land13060881

Casini D., Barsali T., Rizzo A. M. and Chiaramonti D. (2021). Production and characterization of co-composted biochar and digestate from biomass anaerobic digestion. *Biomass Conversion and Biorefinery*, **11**(6), 2271–2279. https://doi.org/10.1007/S13399-019-00482-6

Chakraborty N., Mitra R., Pal S., Ganguly R., Acharya K., Minkina T., Sarkar A. and Keswani C. (2023). Biopesticide consumption in India: insights into the current trends. *Agriculture*, **13**(3), 557. https://doi.org/10.3390/agriculture13030557

Charis G., Patel B., Rashama C. and Nyamunda B. C. (2024). Progress towards the standardization of test parameters and protocols for evaluating densified fuel products. *Paliva*, **16**(1), 16–29, https://doi.org/10.35933/paliva.2024.01.03

Cho C., Park S. and Sung K. (2013). Subsurface and plant contamination during natural attenuation and phytoremediation of silt loam contaminated with chlorinated organic compounds. *Vadose Zone Journal*, **12**(1), 1–11. https://doi.org/10.2136/vzj2012.0081

Coimbra E. C. L. and Borges A. C. (2023). Removing Mn, Cu and Fe from real wastewaters with macrophytes: reviewing the relationship between environmental factors and plants' uptake capacity. *Toxics*, **11**(2), 158. https://doi.org/10.3390/toxics11020158

Cui Y., Zeng Y., Hu H., Zhang Y., Wang D. and Feng D. L. (2024). Biochar, phosphate, and magnesium oxide in seaweed and cornstarch dregs co-composting: enhancing organic matter degradation, humification, and nitrogen retention. *Waste Management*, **187**, 207–217. https://doi.org/10.1016/j.wasman.2024.07.024

Dar N. A., Pandit A. K. and Ganai B. A. (2014). Factors affecting the distribution patterns of aquatic macrophytes. *Limnological Review*, **14**(2), 75–81, https://doi.org/10.2478/limre-2014-0008

Deshmukh U. B., Shende M. B. and Rathor O. S. (2016). Aquatic macrophytes biodiversity assessment from Asolamendha reservoir of Chandrapur district, Maharashtra State (India). *International Journal of Applied Research*, **2**(1), 293–298.

Dijoo Z. K., Ali R. and Hameed M. (2020). Role of free-floating aquatic macrophytes in abatement of the disturbed environs. In: Bhat R. A. and Hakeem K. R. (eds), Bioremediation and Biotechnology, Vol 4: Techniques for Noxious Substances Remediation, Springer International Publishing, Cham, pp. 259–274. https://doi.org/10.1007/978-3-030-48690-7_12

Edgar V. N., Fabián F. L., Mario P. C. J. and Ileana V. R. (2021). Coupling plant biomass derived from phytoremediation of potential toxic-metal-polluted soils to bioenergy production and high-value by-products–A review. *Applied Sciences*, **11**(7), 2982. https://doi.org/10.3390/app11072982

Elkaliny N. E., Alzamel N. M., Moussa S. H., Elodamy N. I., Madkor E. A., Ibrahim E. M., Elshobary M. E. and Ismail G. A. (2024). Macroalgae bioplastics: A sustainable shift to mitigate the ecological impact of petroleum-based plastics. *Polymers*, **16**(9), 1246. https://doi.org/10.3390/polym16091246

Elsalamony D., Hassouna M., Zaghloul T. I., He Z. and Moustafa H. (2024). Bioenergy production from chicken feather waste by anaerobic digestion and bioelectrochemical systems. *Microbial Cell Factories*, **23**, 102. https://doi.org/10.1186/s12934-024-02374-5

Felicien S., Faridullah F., Irshad M., Bacha A., Hafeez F. and Nduwamungu J. (2024). Effect of biochar on composting of cow manure and kitchen waste. *Land*, **13**(10), 1545. https://doi.org/10.3390/land13101545

Feng D., Cui Y., Zeng Y., Wang D., Zhang H., Zhang Y. and Song W. (2024). Enhancing compost quality through biochar and oyster shell amendments in the co-composting of seaweed and sugar residue. *Chemosphere*, **366**, 143500. https://doi.org/10.1016/j.chemosphere.2024.143500

Forgács G., Forgács G., Alinezhad S., Mirabdollah A., Feuk-Lagerstedt E. and Horváth I. S. (2011). Biological treatment of chicken feather waste for improved biogas production. *Journal of Environmental Sciences–China*, **23**(10), 1747–1753. https://doi.org/10.1016/S1001-0742(10)60648-1

Gan L. (2022). Sequential hybrid constructed wetland for effective nutrient removal from municipal wastewater treatment plant effluent: a full scale case study. *Environmental Engineering Science*, **39**(2), 168–177. https://doi.org/10.1089/ees.2021.0041

Harney N. V. (2014). Macrophytes biodiversity of Dudhala lake of Bhadrawati, district Chandrapur (MS), India. *Asian Journal of Multidisciplinary Studies*, **2**(4), 69–72.

He T., Li H., Jiao F., Zhang Y. and He X. (2024). A study of the effect of biochar additive on the manure–compost–soil process and its bacterial succession. *Sustainability*, **16**(18), 7910. https://doi.org/10.3390/su16187910

Ibrahim I., Hassan M. A., Abd-Aziz S., Shirai Y., Andou Y., Othman M. R., Ali A. A. M. and Zakaria M. R. (2017). Reduction of residual pollutants from biologically treated palm oil mill effluent final discharge by steam activated bioadsorbent from oil palm biomass. *Journal of Cleaner Production*, **141**, 122–127. https://doi.org/10.1016/j.jclepro.2016.09.066

İlay R. (2024). Biochar production from various low-cost marine wastes using different production methods: characterization of biochar and marine feedstock for agricultural purposes. *Marine Pollution Bulletin*, **205**, 116623. https://doi.org/10.1016/j.marpolbul.2024.116623

Ingle S. R., Kamble B. M. and Patil V. S. (2024). Biochar: a comprehensive review of production, properties and applications. *International Journal of Advanced Biochemistry Research*, **8**, 21–29. https://doi.org/10.33545/26174693.2024.v8.i1sa.282

Jain C. K., Malik D. S. and Yadav A. K. (2016). Applicability of plant-based biosorbents in the removal of heavy metals: a review. *Environmental Processes*, **3**(2), 495–523. https://doi.org/10.1007/s40710-016-0143-5

Jamil F., Inayat A., Hussain M., Akhter P., Abideen Z., Ghenaï C., Shanableh A. and Abdellatief T. M. (2024). Valorization of waste biomass to biofuels for power production and transportation in optimized way: a comprehensive review. *Advanced Energy and Sustainability Research*, **5**(10), 2400104. https://doi.org/10.1002/aesr.202400104

Jena J. K., Patro B., Patri P., Khuntia C. P., Tripathy N. K., Sinha S., Sarangi N. and Ayyappan S. (2011). Biological treatment of domestic sewage through duckweed–cum–fish culture: a pilot-scale study. *Indian Journal of Fisheries*, **57**(4), 45–52, https://epubs.icar.org.in/index.php/IJF/article/view/6722

Jian M. (2018). Distribution of aquatic macrophytes in the Le'an river and its indicative evaluation on heavy metal pollution. In: Chinese Water Systems: Volume 3: Poyang Lake Basin, Springer International Publishing, Cham, pp. 125–166. https://doi.org/10.1007/978-3-319-97725-6_10

Justin L. D. and Olukanni D. O. (2024). Efficiency evaluation of wastewater treatment by three macrophytes using a pilot-constructed wetland system in Ota, Nigeria. *Journal of Water and Health*, **22**, 2040–2053. https://doi.org/10.2166/wh.2024.116

Kabir M. M. (2015). Bioprocessing of recalcitrant substrates for biogas production. Doctoral dissertation, Högskolan i Borås.

Karnwal A., Martolia S., Dohroo A., Al-Tawaha A. R. M. S. and Malik T. (2024). Exploring bioremediation strategies for heavy metals and POPs pollution: the role of microbes, plants, and nanotechnology. *Frontiers in Environmental Science*, **12**, 1397850. https://doi.org/10.3389/fenvs.2024.1397850

Kaushik A., Priyadharshini P., Manimegalai S., Palaniselvam V. and Parthiban K. T. (2024). Biochar production from plant residues: a sustainable approach for carbon sequestration and soil fertility improvement. *Archives of Current Research International*, **24**(9), 1–13. https://doi.org/10.9734/acri/2024/v24i9864

Khan S., Thaher M., AbdulQuadir M., Faisal M., Mehariya S., Al-Najjar M. A. A., Al-Jabri H. M. S. J. and Das P. (2023). Utilization of microalgae for urban wastewater treatment and valorization of treated wastewater and biomass for biofertilizer applications. *Sustainability*, **15**, 16019. https://doi.org/10.3390/su152216019

Khatun A., Pal S., Mukherjee A. K., Samanta P., Mondal S., Kole D., Chandra P. and Ghosh A. R. (2016). Evaluation of metal contamination and phytoremediation potential of aquatic macrophytes of East Kolkata Wetlands, India. *Environmental Health and Toxicology*, **31**, e2016021. https://doi.org/10.5620/eht.e2016021

Kotamraju A., Logan M. and Lens P. N. (2023). Integrated bioprocess for Se (VI) remediation using duckweed: coupling selenate removal to biogas production. *Journal of Hazardous Materials*, **459**, 132134. https://doi.org/10.1016/j.jhazmat.2023.132134

Kumar J., Kumar P. and Kushwaha R. (2021). Significance of keratinophiles in biofertilizer development from keratinous waste: Upcoming perspective, pp. 95–101. https://doi.org/10.1016/B978-0-12-821667-5.00006-3

Kumar K. A., Subalakshmi R. J., Jayanthi M., Abirami G., Vijayan D., Prabhu S. V. and Baskaran L. (2022). Production and characterization of enriched vermicompost from banana leaf biomass waste activated by biochar integration. *Environmental Research*, **219**, 115090. https://doi.org/10.1016/j.envres.2022.115090

Kumar R., Kundu D. K., Kormoker T., Joshi S., Rose P. K., Kumar S. A., Sahoo P. K., Sharma P. and Lamba J. (2024). Phycoremediation of potentially toxic elements for agricultural and industrial wastewater treatment: recent advances, challenges, and future prospects. *Desalination and Water Treatment*, **319**, 100505. https://doi.org/10.1016/j.dwt.2024.100505

Li C., Wang Y., Xie S., Wang R., Sheng H., Yang H. and Yuan Z. (2024). Synergistic treatment of sewage sludge and food waste digestate residues for efficient energy recovery and biochar preparation by hydrothermal pretreatment, anaerobic digestion, and pyrolysis. *Applied Energy*, **364**, 123203. https://doi.org/10.1016/j.apenergy.2024.123203

Ling X., Dai Y., Tai Y., Jin C., Li Q., Zhang X. and Yang Y. (2025). Aquatic plants mitigate pollution by enhancing the degradation of atrazine and diuron present in agricultural runoff. *International Journal of Phytoremediation*, **27**(6), 751–760. https://doi.org/10.1080/15226514.2024.2442639

Lutzu G. A., Parsaeimehr A., Ozbay G., Ciurli A., Bacci L., Rao A. R., Ravishankar G. A. and Concas A. (2024). Microalgae and cyanobacteria role in sustainable agriculture: from wastewater treatment to biofertilizer production. *Algae Mediated Bioremediation: Industrial Prospectives*, **2**, 565–618. https://doi.org/10.1002/9783527843367.ch29

Mahajan B., Shastri S. and Londhe S. (2023). Waste removal efficiencies of floating macrophytes for restoration of polluted stream: an experimental analysis. *Urban Science*, **7**(1), 27. https://doi.org/10.3390/urbansci7010027

Malcorps W., Kok B., van 't Land M., Fritz M., van Doren D., Servin K., van der Heijden P., Palmer R., Auchterlonie N. A., Rietkerk M., Santos M. J. and Davies S. J. (2019). The sustainability conundrum of fishmeal substitution by plant ingredients in shrimp feeds. *Sustainability*, **11**(4), 1212. https://doi.org/10.3390/su11041212

Mehra V., Chatterjee S., Pai A., Chandrashekar K., Naha A. and B. V. K. (2022). Bioconversion of recalcitrant keratin rich wastes: a pragmatic approach towards sustainable development. *RASAYAN Journal of Chemistry*, **2022**(Special Issue), 36–42. https://doi.org/10.31788/rjc.2022.1558070

Mikajlo I., Lerch T. Z., Louvel B., Hynšt J., Záhora J. and Pourrut B. (2024). Composted biochar versus compost with biochar: effects on soil properties and plant growth. *Biochar*, **6**(1), 85. https://doi.org/10.1007/s42773-024-00379-2

Miranda A. F., Biswas B., Ramkumar N., Singh R., Kumar J., James A., Roddick F., Lal B., Subudhi S., Bhaskar R. and Mouradov A. (2016). Aquatic plant *Azolla* as the universal feedstock for biofuel production. *Biotechnology for Biofuels*, **9**(1), 221. https://doi.org/10.1186/s13068-016-0628-5

Miranda R., Boechat C. L., Bomfim M. R., Santos J. A. G., Coelho D. G., Assunção S. J. R., Cardoso K. M. and Cardoso E. B. (2022). Phytoremediation: a sustainable green approach for environmental cleanup, pp. 49–75. https://doi.org/10.1016/b978-0-323-85763-5.00017-9

Mogany T., Bhola V. and Bux F. (2024). Algal-based bioplastics: global trends in applied research, technologies, and commercialization. *Environmental Science and Pollution Research International*, **31**(26), 38022–38044. https://doi.org/10.1007/s11356-024-33644-9

Mohan A., Bhatt S. M., Girdhar M., Goyal G., Ansari A. A. and Rehman H. (2016). Current technical perspective and application of aquatic weeds in phytoremediation. Springer International Publishing, pp. 269–289. https://doi.org/10.1007/978-3-319-40148-5_9

Mohanty S. S. and Mohanty K. (2023). Valorization of *Chlorella thermophila* biomass cultivated in dairy wastewater for biopesticide production against bacterial rice blight: a circular biorefinery approach. *BMC Plant Biology*, **23**(1), 644. https://doi.org/10.1186/s12870-023-04579-z

Mohanty M., Pattnaik M. M., Mishra A. K. and Patra H. K. (2023). Phytoefficacy of *Eichhornia crassipes* (Mart.) Solms-Laub for aqua-remediation of hexavalent chromium in chromite mine effluent of South Kaliapani, Odisha, India. *Environmental Science and Pollution Research*, **30**(15), 43927–43931. https://doi.org/10.1007/s11356-023-25294-0

Mustafa H. M. and Hayder G. (2020). Performance of *Pistia stratiotes*, *Salvinia molesta*, and *Eichhornia crassipes* aquatic plants in the tertiary treatment of domestic wastewater with varying retention times. *Applied Sciences*, **10**(24), 9105. https://doi.org/10.3390/APP10249105

Nad S., Konar U., Chattaraj S. and Ganguly A. (2024). Valorization of feather waste by microbial enzymatic activity: bioconversion, production and application. In: Saha S. P., Mazumdar D., Roy S. and Mathur P. (eds), Agro-waste to Microbe Assisted Value Added Product: Challenges and Future Prospects, Springer, Cham, pp. 337–363. https://doi.org/10.1007/978-3-031-58025-3

Nafea E. M. A. (2019). Floating macrophytes efficiency for removing of heavy metals and phenol from wastewaters. *Egyptian Journal of Aquatic Biology and Fisheries*, **23**(4), 1–9. https://doi.org/10.21608/EJABF.2019.50020

Narayanasamy A., Patel S. K., Singh N., Rohit M. V. and Lee J. K. (2024). Valorization of algal biomass to produce microbial polyhydroxyalkanoates: recent updates, challenges, and perspectives. *Polymers*, **16**(15), 2227. https://doi.org/10.3390/polym16152227

Nargotra P., Sharma V., Lee Y. C., Tsai Y. H., Liu Y. C., Shieh C. J., Tsai M-L., Dong C-D. and Kuo C. H. (2022). Microbial lignocellulolytic enzymes for the effective valorization of lignocellulosic biomass: a review. *Catalysts*, **13**(1), 83. https://doi.org/10.3390/catal13010083

Navarro-López E., Gallardo-Rodríguez J. J., del Carmen Cerón-García M., Gallego-López I., Acién-Fernández F. G. and Molina-Grima E. (2023). Extraction of phytostimulant molecules from *Scenedesmus almeriensis* using different extractor systems. *Journal of Applied Phycology*, **35**(2), 701–711. https://doi.org/10.1007/s10811-023-02919-y

Nazir S. (2020). The use of *Spirodela polyrrhiza* (duckweed) and *Eichhornia crassipes* (water hyacinth) to phytoremediate wastewater in Guyana. *Asian Journal of Applied Science and Technology*, **4**(3), 69–75. https://doi.org/10.38177/ajast.2020.4310

Noerhayati E., Suprapto B., Rahmawati A. and Mustika S. N. (2023). Improving wastewater quality system using the Internet of Things-based phytoremediation method. *Journal of Ecological Engineering*, **24**(3), 254–262. https://doi.org/10.12911/22998993/158382

Omweno J. O., Argwings O., Sieberi B. M., Angima M. and Ondieki P. M. (2024). Utilization of aquatic plants and microalgae for sustainable aquaculture production and potential biotechnological applications. *Tropical Environment. Biology, and Technology*, **2**(2), 57–71. https://doi.org/10.53623/tebt.v2i2.451

Padme Y. L. (2024). Wastewater management by using effective phytoremediator duckweed. *International Journal of Science and Research Archive*, **13**(2), 2619–2622. https://doi.org/10.30574/ijsra.2024.13.2.2437

Pandey V. C. (2022). Algae and Aquatic Macrophytes in Cities: Bioremediation, Biomass, Biofuels and Bioproducts. Elsevier, Amsterdam.

Pang Y. L., Quek Y. Y., Lim S. and Shuit S. H. (2023). Review on phytoremediation potential of floating aquatic plants for heavy metals: a promising approach. *Sustainability*, **15**(2), 1290. https://doi.org/10.3390/su15021290

Patra S. R., Kabi S. K., Kumar M. and Dhal N. K. (2024). Post-phytoremediation strategies for biomass management toward its sustainable utilization and economic opportunities in the context of heavy metals pollution. In: Kuanar A., Das A. P., Kar D. and Shah M. P. (eds), Sustainable Management of Environmental Pollutants through Phytoremediation, CRC Press, Boca Raton, pp. 1–17.

Patra S., Saikia P. and Kumar A. (2025). Management of aquatic ecosystems and aquatic vegetation for environmental sustainability. In: Aquatic Ecosystems Monitoring, CRC Press, pp. 205–217.

Rahman A. and Miller C. D. (2017). Microalgae as a source of bioplastics. In: Algal Green Chemistry, Elsevier, pp. 121–138. https://doi.org/10.1016/B978-0-444-63784-0.00006-0

Rajput N. A., Laghari M. Y., Mangio H. R. and Soothar R. K. (2024). Enhanced phosphorus and potassium recovery through co-pyrolysis of nutrient-rich biomass feedstocks for engineered biochar production. *Bioresource Technology Reports*, **26**, 101818. https://doi.org/10.1016/j.biteb.2024.101818

Rasool S., Ahmad I., Jamal A., Saeed M. F., Zakir A., Abbas G., Seleiman M. F. and Caballero-Calvo A. (2023). Evaluation of phytoremediation potential of an aquatic macrophyte (*Eichhornia crassipes*) in wastewater treatment. *Sustainability*, **15**(15), 11533. https://doi.org/10.3390/su151511533

Ravindran R. and Jaiswal A. K. (2016). Microbial enzyme production using lignocellulosic food industry wastes as feedstock: a review. *Bioengineering*, **3**(4), 30. https://doi.org/10.3390/bioengineering3040030

Reddy C. C., Khilji I. A., Gupta A., Bhuyar P., Mahmood S., Al-Japairai K. A. S. and Chua G. K. (2021). Valorization of keratin waste biomass and its potential applications. *Journal of Water Process Engineering*, **40**, 101707. https://doi.org/10.1016/J.JWPE.2020.101707

Rejmankova E. (2011). The role of macrophytes in wetland ecosystems. *Journal of Ecology and Environment*, **34**(4), 333–345, https://doi.org/10.5141/JEFB.2011.044

Rodríguez-Palacio M. C., Cabrera-Cruz R. B. E., Rolón-Aguilar J. C., Lozano-Ramírez C., Galeana-Hurtado L. J. and Morales-Avendaño E. D. (2018). Comparative study of the efficiency of removal of N and P of municipal wastewater and leached of vermicomposting with five strains of microalgae. *Desalination and Water Treatment*, **131**, 180–186. https://doi.org/10.5004/dwt.2018.23035

Roy D., Sreekanth D., Pawar D., Mahawar H. and Barman K. K. (2021). Phytoremediation of arsenic contaminated water using aquatic, semi-aquatic and submerged weeds. In: Ferreira Mendes K., Nogueira de Sousa R. and Cabral Mielke K. (eds), Biodegradation Technology of Organic and Inorganic Pollutants. IntechOpen. https://doi.org/10.5772/intechopen.98961

Safafar H. (2017). Micro-algae biomass as an alternative resource for fishmeal and fish oil in the production of fish feed.

Saha A., Mukherjee P., Roy K., Sen K. and Sanyal T. (2022). A review on phyto-remediation by aquatic macrophytes: a natural promising tool for sustainable management of ecosystem. *International Journal of Experimental Research and Review*, **27**, 9–31. https://doi.org/10.52756/ijerr.2022.v27.002

Saidin Z. H., Tajuddin R. M., Kamarun D. and Rahman N. A. (2014). Performance of aquatic macrophytes on removal and accumulation of sulfate and potassium from domestic wastewater. InCIEC 2013: Proceedings of the International Civil and Infrastructure Engineering Conference 2013, pp. 769–781. Singapore: Springer Singapore. https://doi.org/10.1007/978-981-4585-02-6_66

Sanyal P., Chakraborty S. K. and Ghosh P. B. (2015). Phytoremediation of sewage-fed wetlands of East-Kolkata, India: a case study. *International Research Journal of Environment Sciences*, **4**(1), 80–89.

Sekar M., Mathimani T., Alagumalai A., Chi N. T. L., Duc P. A., Bhatia S. K., Brindhadevi K. and Pugazhendhi A. (2021). A review on the pyrolysis of algal biomass for biochar and bio-oil–bottlenecks and scope. *Fuel*, **283**, 119190. https://doi.org/10.1016/j.fuel.2020.119190

Shaika N. A., Khan S., Sultana S., Haque M. M., Rao A. R. and Ravishankar G. A. (2024). Aquaculture wastewater treatment through microalgae for biomass production and its use in agriculture, feed, and energy applications. *Algae Mediated Bioremediation: Industrial Prospectives*, **1**, 295–315. https://doi.org/10.1002/9783527843367.ch14

Sharma P., Pandey A., Tong Y. W. and Ngo H. H. (eds). (2022). Current Developments in Biotechnology and Bioengineering: Advances in Phytoremediation Technology. Elsevier, Amsterdam.

Siddiqui A. J., Jahan S., Adnan M., Ashraf S. A. and Singh R. (2023). Macrophytes and their role in wetland ecosystems. In: Aquatic Macrophytes: Ecology, Functions and Services, Springer Nature Singapore, Singapore, pp. 119–138. https://doi.org/10.1007/978-981-99-3822-3_6

Singh A. and Kumar S. (2022). Application of aquatic plants dead biomass in remediation of heavy metals pollution by adsorption: a review. *Indian Journal of Science and Technology*, **15**(16), 729–735. https://doi.org/10.17485/IJST/v15i16.82

Sinha R., Singh A. K., Bauddh K., Sharma, T. R. and Sharma P. (2021). Phytomining: A sustainable approach for recovery and extraction of valuable metals. In: Bauddh K., Korstad J. and Sharma P. (eds), Phytorestoration of Abandoned Mining and Oil Drilling Sites. Elsevier, pp. 487–506. https://doi.org/10.1016/B978-0-12-821200-4.00013-3

Song U. (2020). Improvement of soil properties and plant responses by compost generated from biomass of phytoremediation plant. *Environmental Engineering Research*, **25**(5), 638–644. https://doi.org/10.4491/eer.2019.59

Song U. and Park H. (2017). Importance of biomass management acts and policies after phytoremediation. *Journal of Ecology and Environment*, **41**(1), 13. https://doi.org/10.1186/s41610-017-0033-4

Song S., Sheng Q., Zhu Z. and Liu Y. (2023). Application of multi-plant symbiotic systems in phytoremediation: a bibliometric review. *Sustainability*, **15**(16), 12252. https://doi.org/10.3390/su151612252

Stegenta-Dąbrowska S., Syguła E., Bednik M. and Rosik J. (2024). Effective carbon dioxide mitigation and improvement of compost nutrients with the use of composts' biochar. *Materials*, **17**, 563. https://doi.org/10.3390/ma17030563

Sun J., Norouzi O. and Mašek O. (2022). A state-of-the-art review on algae pyrolysis for bioenergy and biochar production. *Bioresource Technology*, **346**, 126258. https://doi.org/10.1016/j.biortech.2021.126258

Tang J. (2024). Application of constructed wetlands to treat sewage by the Government in China. *Scientific and Social Research*, **6**(6), 266–269. https://doi.org/10.26689/ssr.v6i6.7414

Tasnuva F. (2025). *Phytoremediation: Exploring opportunity in Bangladesh.*

Thakur T., Barya M., Dutta J., Mukherjee P., Thakur A., Swamy S. and Anderson J. T. (2023). Integrated phytobial remediation of dissolved pollutants from domestic wastewater through constructed wetlands: an interactive macrophyte–microbe-based green and low–cost decontamination technology with prospective resource recovery. *Water*, **15**, 3877. https://doi.org/10.3390/w15223877

Thampatti K. C. M., Beena V., Meera A. V. and Ajayan A. S. (2020). Phytoremediation of metals by aquatic macrophytes, pp. 153–204. https://doi.org/10.1007/978-3-030-00099-8_6

Töre G. Y. and Özkoç Ö. B. (2022). Recent developments in aquatic macrophytes for environmental pollution control: a case study on heavy metal removal from lake water and agricultural return wastewater with the use of duckweed (*Lemnaceae*). In: Phytoremediation Technology for the Removal of Heavy Metals and Other Contaminants From Soil and Water, Elsevier, pp. 75–127. https://doi.org/10.1016/B978-0-323-85763-5.00005-2

Ugya A. Y., Chen H. and Wang Q. (2023). Microalgae biofilm carbon and nitrogen sequestration as a tool for economic and environmental sustainability. *Critical Reviews in Environmental Science and Technology*, **53**(23), 2055–2070. https://doi.org/10.1080/10643389.2023.2209492

Velasco-Arroyo B., Curiel-Alegre S., Khan A. H. A., Rumbo C., Pérez-Alonso D., Rad C., De Wilde H., Pérez-de-Mora A. and Barros R. (2024). Phytostabilization of metal(loid)s by ten emergent macrophytes following a 90-day exposure to industrially contaminated groundwater. *New Biotechnology*, **79**, 50–59. https://doi.org/10.1016/j.nbt.2023.12.003

Waliczek T. M., Drewery M., McMoran A., Eriksen M. and Hale J. (2023). Assessing the quality of compost produced from human hair and pet fur waste. *Compost Science & Utilization*, **29**, 57–64. https://doi.org/10.1080/1065657x.2023.2167749

Wang H., Su Z., Ren S., Zhang P., Li H., Guo X. and Liu L. (2024). Combined use of biochar and microbial agents can promote lignocellulosic degradation microbial community optimization during composting of submerged plants. *Fermentation*, **10**, 70. https://doi.org/10.3390/fermentation10010070

Waqar R., Rahman S., Iqbal J., Kaleem M., Minhas L. A., Ullah N., Kausar F., Chalgham W., Al-Misned F. A., El-Serehy H. A. and Mumtaz A. S. (2023). Biosorption potential of *Desmodesmus* sp. for the sequestration of cadmium and lead from contaminated water. *Sustainability*, **15**(15), 11634. https://doi.org/10.3390/su151511634

Wibowo Y. G., Nugraha A. T. and Rohman A. (2023). Phytoremediation of several wastewater sources using *Pistia stratiotes* and *Eichhornia crassipes* in Indonesia. *Environmental Nanotechnology. Monitoring and Management*, **20**, 100781. https://doi.org/10.1016/j.enmm.2023.100781

Yatoo A. M., Bhat S. A., Ali M. N., Baba Z. A. and Zaheen Z. (2022). Production of nutrient-enriched vermicompost from aquatic macrophytes supplemented with kitchen waste: assessment of nutrient changes, phytotoxicity, and earthworm biodynamics. *Agronomy*, **12**(6), 1303. https://doi.org/10.3390/agronomy12061303

Yin W., Zhang Z., Liu T., Xu J., Xiao S. and Xu Y. (2020). N-doped animal keratin waste porous biochar derived from *Trapa natans* husks. *Materials*, **13**(4), 987. https://doi.org/10.3390/MA13040987

Yongabi K. A., Nagabhatla N. and Soto Rios P. C. (2018). Phytoremediation eco-models using indigenous macrophytes and phytomaterials. Springer, Cham, pp. 253–273. https://doi.org/10.1007/978-3-319-67416-2_9

Yuan Z., Wang Y., Zhu L., Zhang C. and Sun Y. (2024). Machine-learning-aided biochar production from aquatic biomass. *Carbon Research*, **3**(1), 77. https://doi.org/10.1007/s44246-024-00169-2

Zaman W., Ali S. and Akhtar M. S. (2024). Harnessing the power of plants: innovative approaches to pollution prevention and mitigation. *Sustainability*, **16**(23), 10587. https://doi.org/10.3390/su162310587

Zhang P., Zhang T., Zhang J., Liu H., Chicaiza-Ortiz C., Lee J. T., He Y., Dai Y. and Tong Y. W. (2024a). A machine learning assisted prediction of potential biochar and its applications in anaerobic digestion for valuable chemicals and energy recovery from organic waste. *Carbon Neutrality*, **3**, 2. https://doi.org/10.1007/s43979-023-00078-0

Zhang X., Hua J., Zhu Y., Ding X., Zhang Q., Zhang T., Yang D. and Qiu F. (2024b). Birnessite-type MnO_2 modified sustainable biomass fiber toward adsorption removal heavy metal ion from actual river aquatic environment. *Langmuir*, **40**(16), 8738–8750. https://doi.org/10.1021/acs.langmuir.4c00723

Zheng Y., Wang X. C., Dzakpasu M., Ge Y., Zhao Y. and Xiong J. (2016). Performance of a pilot demonstration-scale hybrid constructed wetland system for on-site treatment of polluted urban river water in Northwestern China. *Environmental Science and Pollution Research*, **23**(1), 447–454. https://doi.org/10.1007/S11356-015-5207-Y

Znad H., Awual M. R. and Martini S. (2022). The utilization of algae and seaweed biomass for bioremediation of heavy metal-contaminated wastewater. *Molecules*, **27**(4), 1275. https://doi.org/10.3390/molecules27041275

doi: 10.2166/9781789065404_0093

Chapter 5

Innovative energy recovery in wastewater treatment for sustainable environmental and agricultural resilience

Gourav Sabharwal[1], *Sushmita M. Dadhich*[2]* *and Indrajit Pal*[3]

[1]Department of Agronomy, Tamil Nadu Agricultural University, Coimbatore, India
[2]Division of Soil and Water Conservation Engineering FOAE, SKUAST, Jammu, India
[3]Disaster Preparedness, Mitigation and Management, Asian Institute of Technology, Thailand
**Correspondence: sushmitam52@skuastj.org; sushmita.iitd@gmail.com*

ABSTRACT

The treatment of wastewater is one of the foundations of environmental protection, which is still energy-consuming and resource-intensive. The realization of carbon-neutral operations relies on sustainable transformations in the application of energy recovery technologies. This chapter addresses advanced systems including bioelectrochemical systems, microbial fuel cells, and hydropower integration at wastewater treatment plant outlets, which support reduced fossil fuel dependence and enhanced energy efficiency. It emphasizes the use of life cycle assessment (LCA) as a tool to evaluate environmental impacts and to determine the most efficient recovery pathways. The conversion of organic carbon into valuable products instead of releasing it as greenhouse gases is highlighted as a key innovation. The importance of system dynamics modelling in estimating the global potential for increased treatment coverage through energy recovery is discussed. The feasibility of nutrient cycling, water reuse, and biogas upgrading strategies is analyzed using European case studies and simulation-based models. Finally, the chapter promotes the integration of multiple resource recovery techniques to develop cost-effective and climate-resilient wastewater infrastructure aligned with the sustainable development goals (SDGs).

Keywords: Climate-resilient infrastructure, energy recovery, hydropower turbines, life cycle assessment (LCA), sustainable development goals.

5.1 INTRODUCTION

Water is a major component of human livelihood, it has been utilized for domestic and irrigational purposes. For centuries, raw sewage from human settlements was discharged into nearby natural water bodies due to their inherent ability to dilute and disperse waste. At the time, this practice was considered viable, as the wastewater volumes were low and populations relatively small. Wastewater comprises major percentage of organic and inorganic waste, leads to depletion of oxygen and eutrophication in water bodies (Smith & Schindler, 2009).

Wastewater treatment (WWT) plays a significant role in safeguarding the environment from hazardous pollutants, with an emphasis on effective pollutant removal and resource recovery. As reported by the World Health Organization (WHO), 55.5% of global domestic wastewater was safely treated and converted to potable water in 2020 (Okadera *et al.*, 2020). The ultimate goal behind this is managing pollutants and protection of the environment appropriate with economic, social, political, and health concerns. Hazardous substances like dissolved organic pollutants need to be reduced, particularly toxins in industrial effluent, as they have a major impact on ecosystems. Four billion of the global population suffer from freshwater scarcity, which is threatening water security (Mekonnen & Hoekstra, 2016; Musie & Gonfa, 2023). There is an urgent need for the development of centralized infrastructure that harvests, stores, treats, and transports water for municipal and agricultural uses.

Conventionally, waste from water is disposed of via burning, landfilling or ocean disposal, as well as reused as soil conditioner in agriculture. Conventional treatment infrastructure is energy intensive and harmful for the ecosystem. In wastewater management, energy consumption is more than 2% of total electrical energy available worldwide, whereas some treatment plants exhaust around 20% of a municipality's electricity energy (Singh & Kansal, 2018). In many low- and middle-income countries, wastewater collection and treatment infrastructure remains inadequate, resulting in a significant share of domestic and industrial wastewater being discharged untreated or only minimally treated. Innovative energy recovery in WWT aims to prevent soil and groundwater pollution while reducing energy consumption and promoting recycling and reuse. This chapter evaluates how resource recovery approaches such as microbial fuel cells (MFCs), hydropower integration, nutrient cycling, and biogas upgrading can be incorporated into WWT systems to enhance energy efficiency, support circular management, and expand treatment coverage. WWT inherently relies on physicochemical and microbial processes to separate complex organic and inorganic components from wastewater. These processes significantly reduce the toxicity associated with raw sewage. Prior to wastewater discharge, the removal of total soluble solids (TSS) is critical to protect freshwater clarity and quality (Sangamnere *et al.*, 2023). This step also lowers the overall pollution potential of untreated wastewater.

Conventional WWT facilities were originally designed with limited consideration for energy consumption (Tsalas *et al.*, 2024). However, increasing urbanization has intensified the dependence of WWT plants on energy, making them major energy consumers (Walker *et al.*, 2021). In response, WWT has experienced a major shift toward integrated systems that emphasize resource recovery and net-zero carbon emissions. This transition is particularly relevant for addressing greenhouse gas (GHG) emissions through energy-saving policies and the adoption of innovative technologies that enhance energy recovery while mitigating climate impacts.

In recent years, the wastewater management paradigm has shifted from pollutant removal toward resource recovery, driven by the urgent need to address climate change and reduce reliance on fossil fuels. Wastewater is increasingly viewed not as a burden, but as a renewable resource capable of generating energy, recovering nutrients, and supplying clean water. Advances in bioelectrochemical systems (BES), anaerobic digestion (AD), and hydropower integration enable treatment plants to function as decentralized energy hubs. These solutions are particularly valuable in regions facing energy shortages and water scarcity, offering both environmental and socio-economic benefits. Moreover, integrating circular economy principles into WWT aligns with global frameworks such as the sustainable development goals (SDGs), including SDG 6 (Clean Water and Sanitation), SDG 7 (Affordable and Clean Energy), and SDG 13 (Climate Action). By embedding energy recovery technologies within urban and agricultural systems, countries can strengthen resilience to climate-related disruptions while promoting inclusive development. The convergence of environmental engineering, agricultural sciences, and renewable energy research creates opportunities for interdisciplinary collaboration to develop scalable solutions that address both local and global challenges.

5.2 MATERIALS AND METHODS APPROACH

Secondary data were obtained from Scopus, focusing on key global research about innovative energy recovery in WWT for sustainable environmental and agricultural resilience. A total of 485 documents were identified from year 2015 to 2025 with the search for Article Title, Abstract, Keywords. Further, various studies were reviewed to gain insights into related areas such as the 'energy recovery', 'wastewater treatment,' and 'sustainable agriculture' in addressing these issues (Figure 5.1).

It reveals that between 2015 and 2024 in the field of energy recovery and WWT for sustainable environmental and agricultural resilience there were an average of 52 publications per year. The projected annual growth rate is 15.3%. The trend indicates a slight increase in the number of articles per year: 2015 had the fewest articles and then gradually increasing each year with the highest number of articles in 2017. A higher rise was experienced in 2018 with a total of 36 publications, remaining at 36 or 37 between 2019. After 2020, there was rapid growth in the number of research activities with 65 in 2020 and 63 in 2021. It shows that the world has been paying more attention to sustainability and resource recovery. There was a temporary dip in 2022 (53 articles) which may be as a result of stages of perfection of approaches, and the field rose to the pinnacle in 2023 with 79 publications. There was a slight downturn to 67 articles in 2024, which signals the entry of the stabilization phase since the discipline has matured. It proves the increasing awareness of WWT systems as energy-generating operations instead of energy-gobbling systems.

The inclination can be mainly due to the incorporation of circular economy ideologies that promote the creation of AD, MFCs, and biogas recovery. The present research focuses on nutrient recovery in precision agriculture, intelligent energy recovery, and the power of AI and Internet of Things (IoT),

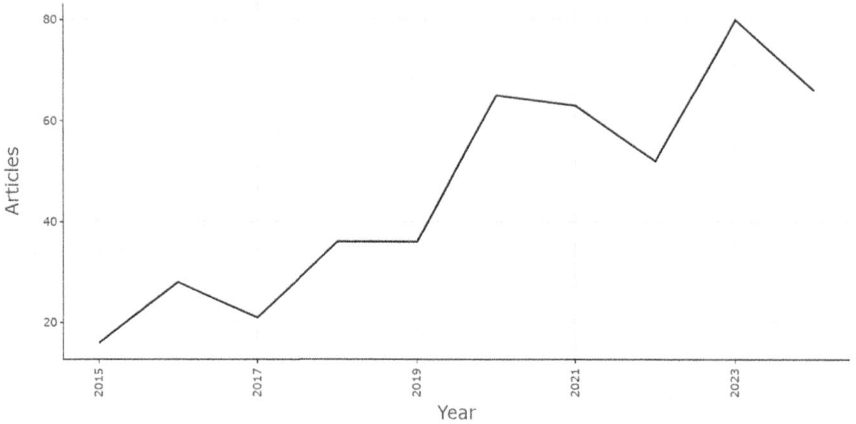

Figure 5.1 Annual number of publications in the field of energy recovery in WWT from 2015 to 2024.

hybrid systems (WWT and renewable energy production), for sanitation, crop or livelihood purposes. Climate change mitigation, food security, and sustainable resource management are emerging key concerns due to the integration of environmental engineering, agricultural sciences, and renewable energy research disciplines, which have together resulted in an interdisciplinary collaboration among these fields.

Figure 5.2 highlights the leading contributors to research on 'Innovative Energy Recovery in Wastewater Treatment for Sustainable Environmental and Agricultural Resilience' (2015–2024). Liu Y tops the list with 14 publications (13.5%). Capodaglio AG and Ferrer J follow with 11 each (10.6%). Li J, Seco A, and Zhang Y contribute 10 each (9.6%), Callegari A adds 9 (8.7%), Wang X and Wang Z produce 8 each (7.7%), and Guo W contributes 7 (6.7%). This distribution highlights that there is an international, interdisciplinary area of focus centered on specialized teams that are progressing MFCs, maximization of AD and integrated agricultural applications. Despite the fast developments, world participation is still disproportionate, particularly in developing areas, suggesting the need to enhance wider collaboration, capacity development, and transfer of knowledge to fast-track the world toward the 2030s.

Figure 5.3 shows that the number of publications in the area of research energy recovery in WWT in various sources has increased throughout the years 2015–2024. *Bioresource Technology* is the most topical source that demonstrates stable growth and is likely in 2024 to occupy the leading position with the largest number of cumulative publications. *Water Research* comes right behind with a good and stable contribution over the decade. *Water (Switzerland)* also shows a steep increase in 2019, with similar values in the last years. *Water Science and Technology* presents moderate but stable contributions, and *Energies* presents a later but a quick growth period starting in 2020 due to its growing relevance to the field. This distribution emphasizes the fact that traditional water and environmental journals continue to play an important role in the field, but

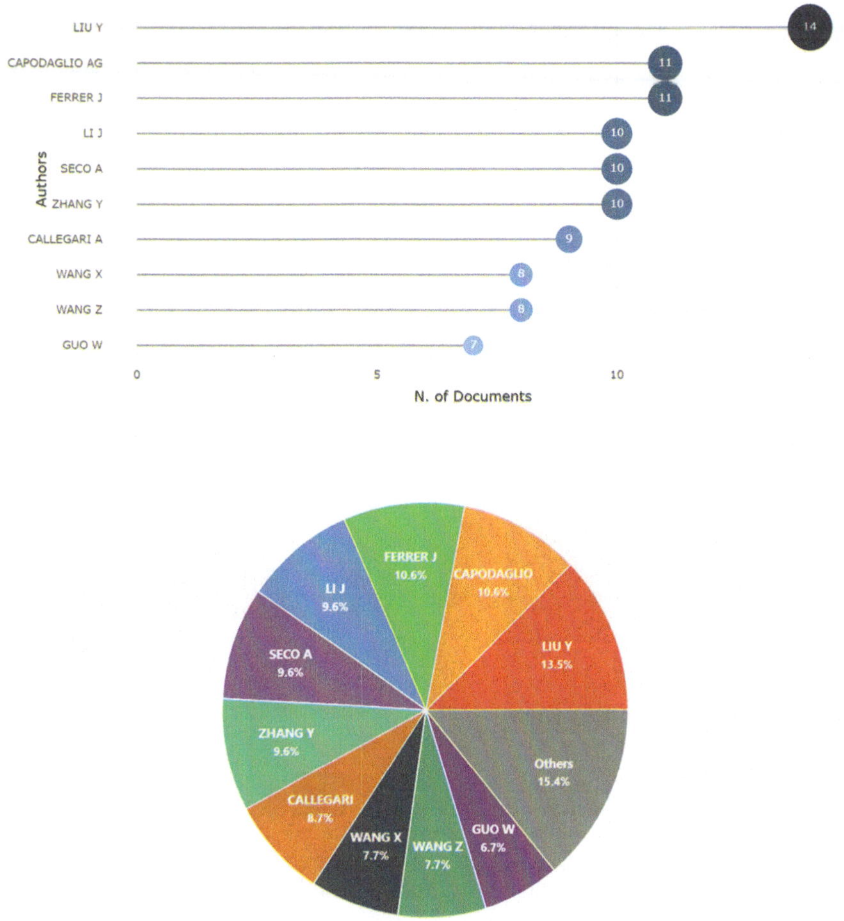

Figure 5.2 Most relevant authors in field of energy recovery in WWT from 2015 to 2024.

interdisciplinary journals like *Energies* are becoming an important input, and this demonstrates the growing research base and increased collaboration across disciplines in renewable energy and environmental engineering.

According to the visualizations, namely the tree map (Figure 5.4) and the word cloud (Figure 5.5), it can be seen that the most important keyword in the research about innovative energy recovery in WWT during the period 2015–2024 is energy recovery, as it is now viewed as a vital source of renewable energy. Other significant keywords are WWT, AD, biogas, MFC, and circular economy, which reflect the focus of the field on advanced treatment technologies, resource recovery, and sustainability. These highlight the interdependence of research subject areas, with environmental resilience and energy production technologies integration being the key themes.

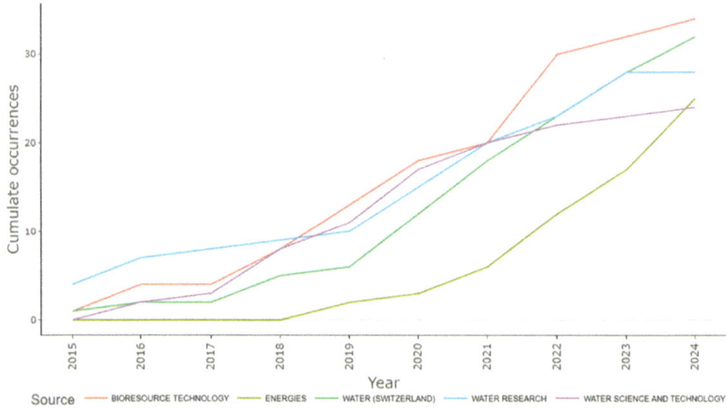

Figure 5.3 Number of documents from different sources in the research domain of energy recovery in WWT from 2015 to 2024.

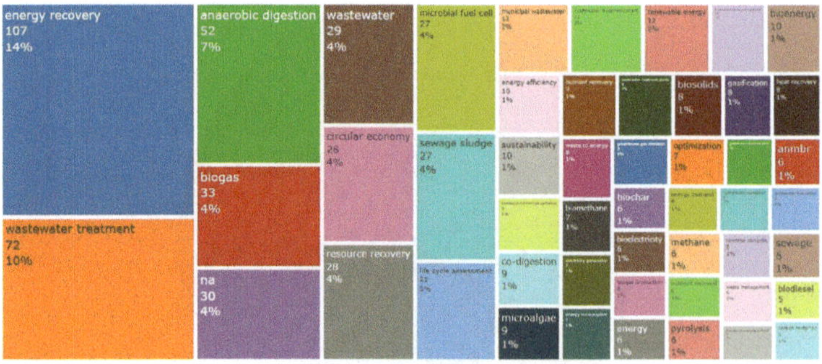

Figure 5.4 TreeMap of mostly used words in the research domain of energy recovery in WWT from 2015 to 2024 (generated via Biblioshiny by authors).

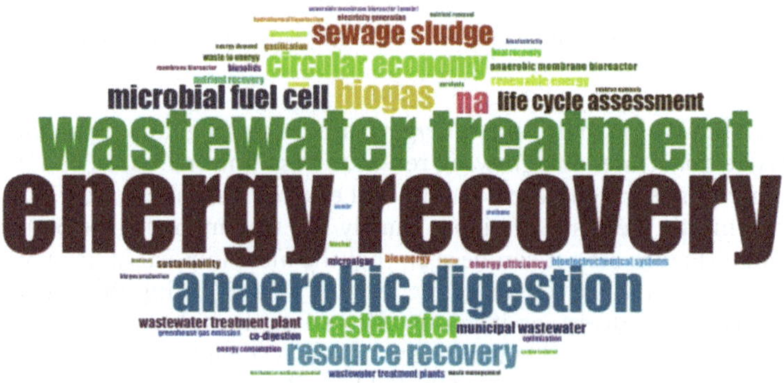

Figure 5.5 Word cloud of most used words in the research domain of energy recovery in WWT from 2015 to 2024 (generated via Biblioshiny by authors).

5.3 RESULTS AND DISCUSSION

The integration of advanced technologies into WWT plants has become a pivotal strategy for energy recovery facility. Energy and resource recovery facilities contribute directly to climate neutrality and decarbonization efforts, as summarized in Table 5.1. Societal acceptance of wastewater reuse is a key factor in accelerating the transition toward circular systems, creating significant opportunities to expand water reuse as an alternative supply source. In recent years, increased efforts have focused on reducing WWT costs while improving net profitability through resource recovery and the reuse of treated wastewater (Mara, 2013; Mateo-Sagasta *et al.*, 2015; McCarty *et al.*, 2011). Resource recovery can be broadly categorized into wastewater reuse, nutrient recovery, and energy recovery. Among these, energy recovery plays a critical role in lowering operational costs and can potentially transform treatment plants into net energy producers (McCarty *et al.*, 2011). Major wastewater resource recovery approaches include biogas production through AD of fecal sludge during organic treatment and the incineration of processed biosolids.

Biogas is generated via the methanogenesis process during AD of organic matter, and comprises methane (60%), carbon dioxide, trace quantity of siloxanes and hydrogen sulfides (Stillwell *et al.*, 2010). Further processing of this can be done by removal of hydrogen sulfides, carbon dioxide, and others for renewable natural gas production (methane – 97%) (Gebrezgabher *et al.*, 2015; Lassner, 2011). Renewable natural gas provides economic benefits as it can be combusted, stored or sold, and even provides benefits of carbon credits (Govindarajan & Elmi, 2013). In aerobic digesters, biosolids are produced as by-products of the digestion process; this nutrient-rich material can be dewatered and incinerated to generate thermal and electrical energy.

5.3.1 Bioelectrochemical systems

BES represent an emerging class of technologies that enable simultaneous WWT and recovery of valuable resources such as energy, nutrients, and metals through microbe–electrode interactions. BES mainly include MFCs, microbial electrolysis cells (MECs), and microbial desalination cells (MDCs), which are often praised for their energy efficiency; however, their overall net energy gains are frequently constrained by high internal resistance, membrane fouling, and operational energy requirements (Meena *et al.*, 2019; Zhang *et al.*, 2022).

Among emerging energy recovery technologies, BES are considered leading innovations with the potential to transform WWT plants into energy-producing facilities. MFCs enable the direct conversion of wastewater into electricity by harnessing the metabolic activity of electrogenic microorganisms, offering a sustainable pathway for energy recovery and resource management. Hybrid system configurations are increasingly explored to enhance both treatment efficiency and energy recovery. Integrating BES with AD, constructed wetlands, and advanced oxidation processes has demonstrated synergistic benefits, including improved removal of organic and inorganic pollutants alongside increased energy production (Tang *et al.*, 2019; Twi-Yeboah *et al.*, 2024). Osmotic membrane technologies have also gained attention due to their

Table 5.1 Advanced technologies for recovering energy from WWT plants.

S. No.	Aim	Interpretation	References
1	Evaluate wastewater energy recovery capacity	Thermal energy recovered from sequencing batch biofilter granular reactor (SBBGR): 0–14.5 kWh, depending on environmental temperature and set point	Piergrossi *et al.* (2018)
2	Assessing the energy recovery and saving potential of MWT technologies	An AD for methane production is most efficient method for chemical energy recovery. Water source heat pumps identified as an effective energy-saving approach for substantial energy savings.	Gong *et al.* (2025)
3	Emerging low energy or energy positive WWT bioprocesses	A paradigm shift from necessary energy inputs and potential energy or resource outputs, from WWT and energy recovery approach	Gao *et al.* (2014)
4	Water in circular economy and resilience (WICER) framework.	(1) Deliver resilient and inclusive water services (2) Design out waste and pollution (3) Preserve and regenerate natural systems	Delgado *et al.* (2024)
5	AD	Producing biogas (mainly methane) that can cover up to half of a plant's energy needs, though conversion efficiency is typically 30–60%	Capodaglio and Olsson (2019)
6	Co-digestion	Supplemental organic waste and CHP systems can further enhance energy yields and move plants toward energy-neutral or energy-positive operation	Sarpong and Gude (2021)
7	MFCs systems	Emerging technologies for direct electricity generation, but large-scale application remains limited by technical challenges	Cecconet *et al.* (2018), Malik *et al.* (2023)
8	Heat exchangers and heat pumps	Energy extraction for district heating, building heating/cooling, or process needs, with studies showing up to 59% reduction in building energy consumption when using wastewater heat	Nagpal *et al.* (2021)
9	Sludge-to-energy technologies	Combustion, pyrolysis, and gasification of sewage sludge are viable for energy recovery, with calorific values comparable to lignite and biomass, but require careful management of pollutants and operational challenges	Oladejo *et al.* (2018), Sarpong and Gude (2021)

ability to harvest osmotic energy while enhancing nutrient and metal recovery, although challenges related to membrane cost and durability persist (Gong *et al.*, 2025). Life cycle assessment (LCA) studies reveal that although BES typically contribute only about 0.01–2% of a system's total energy balance in terms of net electricity production, benefits such as reduced sludge generation, lower GHG emissions, and recovery of valuable by-products can offset these limitations when integrated into broader treatment systems (Malik *et al.*, 2023).

Despite their potential, BES technologies face notable operational and economic challenges, including high internal resistance, electrode degradation, membrane fouling, and the need for low-cost, long-lasting materials, which restrict large-scale implementation (Lima *et al.*, 2024). In addition, system maintenance, external power inputs (particularly for MECs), and pumping requirements may consume more energy than is generated, challenging the perception of BES as inherently energy-positive technologies (Tang *et al.*, 2019). Nevertheless, ongoing research on modular system designs, pilot-scale demonstrations, and the use of waste-derived materials for electrodes is helping to bridge the gap between laboratory-scale success and real-world application (Malik *et al.*, 2023; Twi-Yeboah *et al.*, 2024). Being a part of a wider shift toward the realm of the circular economy in wastewater management, BES are a promising and dynamic solution that would have to be further innovated, characterized by positive policy frameworks, and cross-sectoral integration in order to maximize them.

5.3.2 Anaerobic digestion and biogas optimization

AD is an organically established biological mechanism that breaks down organic substances in the absence of oxygen to produce biogas, which is mainly methane (CH_4) and carbon dioxide (CO_2) with small amounts of hydrogen sulfide (H_2S) and ammonia (NH_3). Raw biogas is typically an unrefined mixture containing about 55–65% methane, with the remaining fraction composed mainly of carbon dioxide and small amounts of other gases. The performance of AD systems is strongly influenced by several factors, including feedstock composition, operational conditions, and the diversity and functionality of the microbial consortia involved.

Different organic substrates exhibit varying biochemical methane potentials (BMPs), which directly affect methane yield. Highly biodegradable substrates such as food waste can generate methane-rich biogas with concentrations reaching up to 60%. When used in co-digestion, particularly with sewage sludge or agricultural residues, food waste can significantly enhance methane production due to synergistic microbial interactions and improved nutrient balance (Malik *et al.*, 2023). Operational conditions such as mesophilic (35–40°C) or thermophilic (50–55°C) temperatures, neutral pH, adequate hydraulic retention time (HRT), and effective volatile solids reduction are critical for maximizing methane production and ensuring process stability (Lima *et al.*, 2024).

The microbial ecology of methanogens, especially acetoclastic and hydrogenotrophic archaea, plays a central role in methane formation. Advances in metagenomics and metabolic modeling have enabled targeted manipulation of microbial communities to improve methane yields and enhance reactor resilience. Studies have shown that strategies such as trace metal supplementation (e.g., nickel and cobalt) and dynamic feeding regimes can substantially increase methane production.

5.3.3 Biogas upgrading and grid integration

To utilize biogas as a renewable alternative to natural gas–particularly for grid injection, transport fuel, or industrial heating–the raw biogas must be upgraded to achieve methane concentrations above 90–97%, producing biomethane. The upgrading process involves removing carbon dioxide, hydrogen sulfide, water vapor, and other trace contaminants to meet quality and safety standards (Adnan *et al.*, 2019; Awe *et al.*, 2017).

Physico-chemical upgrading processes control the business environment and comprise pressure swing adsorption (PSA):

- It involves the use of porous materials such as carbon molecular sieves (CMS-3K) to selectively adsorb CO_2 at pressured conditions. Purity of biomethane in industrial-scale PSA systems has reached 971% and a recovery rate of 862, which is the level of natural gas grids (Liu *et al.*, 2020).
- Water and chemical scrubbing remove CO_2 through solubility differences or chemical reactions (e.g., with mono ethanol amine). Although effective, these methods can be energy-intensive and prone to solvent degradation.
- Membrane separation: it has a high selectivity and can be scaled in a modular manner, although it faces the problem of methane slip and the loss of CH_4 during CO_2 removal where the use of methane recovery units will reduce GHG emissions.
- Biological upgrading is a new promising technology that involves the use of hydrogenotrophic methanogens that are able to reduce CO_2 to methane by injecting hydrogen (H_2) into the biogas stream. This process can deliver in situ or ex situ upgrading, and in ideal cases, the concentrations of methane are above 98%.
- The biological pathway is currently gaining traction because of its low energy needs, its ability to utilize renewable hydrogen obtained by excess wind power or other solar power, and how power-to-gas strategies can be implemented. Nevertheless, it is very important to have accurate control over the dosing of H_2, its mixing, and pH to prevent microbial inhibition and reactor instability.
- Grid integration: to be compatible with natural gas grids, upgraded biomethane must meet stringent standards for methane content ($\geq$95%), calorific value, and impurity limits. Once purified, biomethane is fully interchangeable with fossil natural gas and can be used for residential heating, vehicle fueling (as bio-CNG), or electricity generation.

5.3.4 Hydropower integration at effluent discharge points

Hydropower integration at effluent discharge points is an emerging approach to recover renewable energy from gravity-fed wastewater systems. By harnessing the kinetic and potential energy of treated effluent, WWT plants can generate electricity, reduce operational costs, and contribute to sustainability goals.

Many WWT plants discharge treated effluent by gravity, creating a continuous flow with a measurable elevation drop (head). This flow energy can be captured using low- head hydro turbines such as Kaplan, Archimedes screw, or cross-flow turbines, which are well- suited to the relatively low head (1–3 m) and high flow rates typical of these systems. For example, a Wisconsin WWT plant with a 3-m head and a flow rate of 190 million gallons per day successfully implemented a hubless rim-drive Kaplan turbine, generating an estimated 1564 MWh/year in energy savings (Abbas *et al.*, 2019). Similarly, studies have shown that even small heads (as low as 15 m) can yield significant power when paired with appropriate turbine technology, making retrofitting feasible for many existing plants (Darries *et al.*, 2022).

Several real-world applications demonstrate the viability and benefits of micro-hydro integration at effluent discharge points:

- In Palermo, Italy, a WWT plant compared Kaplan, hydrostatic pressure machine, and cross-flow turbines for a 3-m head, finding that each technology had unique advantages in efficiency and cost-effectiveness, with the Kaplan turbine offering the highest power output (Sinagra *et al.*, 2022).
- In South Africa, a study found that installing an Archimedes screw turbine at a plant's outflow could generate between 79 and 98 kW, outperforming river-based installations and highlighting the potential for widespread retrofitting (Darries *et al.*, 2022).
- In Ireland and the UK, an evaluation of over 100 plants developed a method to estimate power output and payback periods, showing that on-site use of generated electricity often results in shorter payback times than selling to the grid (Power *et al.*, 2014).
- In Malaysia, feasibility studies identified sewage treatment plants with high continuous effluent flow as prime candidates for micro-hydropower, emphasizing the importance of flow rate and head in site selection (Jalil *et al.*, 2019).

5.3.5 Circular carbon pathways: from organic waste to value-added products

The circular carbon pathway reconverts organic waste materials into high value products, which both minimizes the impact on the environment and contributes to making the switch toward sustainable, circular bioeconomy. Biotechnological and thermochemical treatment of diverse organic wastes enables the production of biofuels, biochar, and bioplastics from municipal solid waste, agricultural residues, and food waste. Technologies such as AD, fermentation, pyrolysis, and gasification allow the conversion of these waste streams into biogas, bioethanol, biodiesel, and syngas. For example, the organic fraction of municipal solid waste can be utilized to produce sustainable biofuels and bioplastics, reducing dependence on fossil-based materials and supporting environmentally friendly alternatives (Ebrahimian *et al.*, 2020). Within the circular carbon framework, biochar represents a valuable product generated through pyrolysis. Biochar has multifunctional applications, including soil enhancement, carbon sequestration, and use as a catalyst in further biomass conversion processes. Studies have demonstrated the potential of biochar as a catalyst in thermochemical reactions, with research focusing on improving efficiency and scalability (Li *et al.*, 2025). Similarly, microbial fermentation of organic waste can produce biodegradable plastics such as polyhydroxyalkanoates (PHAs), offering renewable and compostable alternatives to conventional petroleum-based plastics (Ali *et al.*, 2024).

Integrating these conversion technologies into wastewater and solid waste management systems reflects the core principles of a circular bioeconomy. In this model, waste streams are transformed into valuable resources, closing material loops and generating both environmental and economic benefits. Advances in integrated biorefineries, decentralized waste valorization, and hybrid conversion systems further enhance the efficiency and feasibility of circular bioeconomy approaches (Chavan *et al.*, 2022).

5.3.6 Environmental assessment and decision support

Decision-support tools and environmental assessments are essential for advancing waste recovery as a sustainable energy solution. LCA is a key analytical method used to systematically evaluate the environmental impacts of recovery technologies across their entire life cycle, from raw material extraction to final disposal. LCA has been widely applied to compare waste-to-energy options such as incineration, AD, landfill gas recovery, and emerging biorefinery systems. By identifying environmental hotspots and evaluating trade-offs–such as GHG emissions versus energy and resource savings–LCA supports informed comparisons between alternative technologies. Studies indicate that biogas-based combined heat and power (CHP) systems can offer greater net environmental benefits than other recovery options; however, outcomes are highly sensitive to system design, waste characteristics, and the type of energy displaced (Ajanovic, 2013; Chen *et al.*, 2009; Tonini *et al.*, 2012).

To complement the environmental focus of LCA, multi-criteria decision analysis (MCDA) incorporates economic and social dimensions into the evaluation process. MCDA assists decision-makers in balancing environmental impacts, economic feasibility, social acceptance, and regulatory compliance within a transparent and structured framework. The system boundaries used in the two approaches, LCA and MCDA, have a significant impact; depending on the scope of the boundaries used, broader system boundaries may capture avoided effects, and narrower definition systems may miss the indirect effects and long-term outcomes (Finnveden *et al.*, 2009). The integration of LCA and MCDA has thus emerged as a powerful decision-support approach, facilitating context-sensitive planning and optimized resource recovery (Zaman & Lehmann, 2013).

Comparative sustainability analysis also shows that integrated waste management systems, particularly those that focus on material recovery with subsequent energy conversion, generally have a higher environmental performance than the landfill-only or incineration-based systems. Specifically, biorefinery models, yielding a collection of products (e.g., biofuels, electricity, and high-value chemicals), tend to be less ecologically toxic and more resource-efficient than those with a single output (e.g., incineration or standalone AD)

(Nizami *et al.*, 2017). However, the sustainability of these options, on the whole, is also vulnerable to the fluctuations in the feedstock, the level of technological development, and local energy grids, which highlights the necessity to provide dynamic, location-specific considerations that would correspond to the real-life complexities.

5.4 POLICY, FINANCING, AND IMPLEMENTATION PATHWAYS

Technological readiness is not enough to successfully convert BES prototypes into practical use, but also powerful policy backing, innovative financing models and effective institutional structures are needed. These are the pillars of implementation plans that help in bridging the gap between the scientific research and all-scaled and sustainable implementation.

The facilitating regulatory environment is the prerequisite of innovation in BES and other waste-to-energy technologies. Clear and transparent policies offer clarity regarding project acceptance, environmental safety and norms that are vital in bringing in private investment and also dealing with long-term risk. Regulatory certainty also guarantees that the terms of operations of the public–private partnerships (PPP) are done in a fair way with enforceable terms, reducing the concern about opportunistic behavior and equitable cost and profit sharing (Sozen & Shaw, 2002). Furthermore, policies should be in place to control the whole lifecycle of BES projects including design, construction, operation, and transfer in order to uphold the interest and system integrity of the people. Whole-life governance models (Pant *et al.*, 2011), assist in predicting risks at every step of a PPP and advance the social welfare in the form of feedback and adaptive controls. This is particularly relevant for emerging technologies such as BES, where uncertainty remains high and long-term performance is difficult to predict.

Governments can play a critical role in promoting BES adoption by offering targeted financial incentives, including seed funding, low-interest loans, performance-based subsidies, and tax exemptions for research and development equipment. Such policy instruments help reduce investment risks and encourage greater participation from the private sector (Jiang *et al.*, 2023). For example, watershed ecological PPP projects in China have shown that combining stringent environmental regulations with performance-based financial mechanisms can foster sustainable collaboration among multiple stakeholders. PPP have proven particularly effective in innovation-driven sectors by enabling knowledge sharing, risk distribution, and faster commercialization of emerging technologies. However, successful PPPs require more than financial support; they depend on well-aligned incentives, strong institutional frameworks, and the careful selection of strategic partners. Sikombe *et al.* (2024) have discussed that the most effective way to build trust and long-term cooperation in the research-intensive fields is knowledge-sharing platforms, innovation clusters, and well-defined intellectual property regimes. Viability can be further improved by business models that combine the principles of the circular economy when BES produces electricity, clean water, and fertilizer, monetizing them. Involving co-benefits into revenue models, carbon credits or savings on WWT costs, BES projects can attract a larger range of investors and policymakers. The governance structures should be made in a way that is moderate to the populace as well as the efficiency of the structures. Long-term BES projects can be sabotaged by challenges such as lack of cohesion of authority between different jurisdictions, inadequacy of capacity in local institutions, and political instability.

As highlighted by Liu *et al.* (2023), PPPs may contribute to innovation in the design and bidding of a project, but their success after the contract is usually determined by the presence of flexible legal frameworks and continued institutional support. This means that transparency in the process of selecting partners, consulting with the populace, and monitoring performance is vital in gaining the trust of the people and making the project sustainable.

Moreover, coherence in policy in areas of energy, water, agriculture and climate is critical. BES technologies lie between several policy areas, and a lack of alignment can halt the implementation. As technologies grow and mature, responsive decision-making can be facilitated by a supportive inter-agency governance model along with adaptive policy instruments.

5.5 CONCLUSIONS AND RECOMMENDATIONS

The global wastewater industry is at a crossroad where linear solutions of WWT and disposal are shifting into circular ones with priorities of resource recovery and climate change resilience, due to climate change, aging infrastructure, and strain on the available resources. Modern WWT facilities are gradually evolving into versatile hubs where clean water is recovered, renewable energy is generated through biogas and bio electrochemical methods, and valuable nutrients and materials like bioplastics are extracted. These advancements increase the prevalence of energy-self-sufficient operations and cut GHG emissions, while decentralized and modular technologies–such as anaerobic membrane bioreactors and constructed wetlands–improve local system sustainability and adaptability. Durable designs must withstand varying conditions, including temperature shifts, intense rainfall, and rising sea levels, by incorporating robust measures and flexible frameworks. This goal can be realized via innovations in future process integration, climate resilience, and economic viability, supported by targeted demonstration initiatives. Furthermore, appropriate policy structures, financial incentives, and collaborative governance approaches are essential to speed up the adoption of climate-proof, circular wastewater systems that not only process wastewater but also contribute to urban sustainability and resource security goals.

5.5.1 Recommendations

- Decentralization and modularity: promote the implementation of decentralized WWT facilities equipped with energy recovery components such as anaerobic membrane bioreactors and MFCs. These setups offer advantages like expandability, resilience to climate fluctuations, and suitability for peri-urban and rural areas where centralized systems are impractical.

- Water budgeting: develop location-tailored models that integrate energy recovery (for instance, biogas and electricity) with intelligent irrigation scheduling and automated harvesting technologies. This combination can boost water usage efficiency and decrease fossil fuel dependency in farming, notably in water-deficient locations like Jammu.
- Create incentives to utilities based on performance: establish financial mechanisms to reward wastewater utilities for achieving energy-neutral or energy-positive operations. These could include carbon credits, tariffs for feeding renewable electricity on-site into the grid, and subsidies for nutrient recovery technologies.
- Empower capacity building and knowledge transfer: encourage institutional education and technical exchange among organizations in innovation-rich regions and those in developing contexts. This approach will accelerate the adoption of advanced technologies such as BES and hydropower upgrades, while fostering local innovation networks.
- Mandate LCA in infrastructure planning: include LCA and multi-criteria decision analysis (MCDA) as standard components of wastewater infrastructure proposals to ensure sustainability in environmental, economic, and social aspects. This will aid policymakers and engineers in selecting the best technology combinations and avoiding unintended trade-offs.
- Support cross-sector governance models: encourage inter-agency collaboration and synergy among water, energy, agriculture, and climate departments to enhance policy implementation and eliminate regulatory hurdles. A cohesive governance approach can enable improved planning and faster deployment of circular wastewater systems.

REFERENCES

Abbas A. I., Qandil M. D., Al-Haddad M. R., Saravani M. S. and Amano R. S. (2019). Utilization of hydroturbines in wastewater treatment plants. *Journal of Energy Resources Technology*, **141**(6), 062011, https://doi.org/10.1115/1.4042969

Adnan A. I., Ong M. Y., Nomanbhay S., Chew K. W. and Show P. L. (2019). Technologies for biogas upgrading to biomethane: a review. *Bioengineering*, **6**(3), 92, https://doi.org/10.3390/bioengineering6040092

Ajanovic A. (2013). Renewable fuels–A comparative assessment from economic, energetic and ecological point-of-view up to 2050 in EU-countries. *Renewable Energy*, **60**, 733–738, https://doi.org/10.1016/j.renene.2013.06.012

Ali Z., Abdullah M., Yasin M. T., Amanat K., Ahmad K., Ahmed I., Qaisrani M. M. and Khan J. (2024). Organic waste-to-bioplastics: conversion with eco-friendly technologies and approaches for sustainable environment. *Environmental Research*, **244**, 117949, https://doi.org/10.1016/j.envres.2023.117949

Awe O. W., Zhao Y., Nzihou A., Minh D. P. and Lyczko N. (2017). A review of biogas utilisation, purification and upgrading technologies. *Waste and Biomass Valorization*, **8**(2), 267–283, https://doi.org/10.1007/s12649-016-9826-4

Capodaglio A. G. and Olsson G. (2019). Energy issues in sustainable urban wastewater management: use, demand reduction and recovery in the urban water cycle. *Sustainability*, **12**(1), 266, https://doi.org/10.3390/su12010266

Cecconet D., Callegari A., Capodaglio A. G. and Carucci A. (2018). Agro-food industry wastewater treatment with microbial fuel cells: energetic recovery issues. *International Journal of Hydrogen Energy*, **43**(1), 500–511, https://doi.org/10.1016/j.ijhydene.2017.07.231

Chavan S., Bhosale R. R., Gawande P. M. and Shinde S. (2022). Bioconversion of organic wastes into value-added products: a review. *Bioresource Technology*, **344**, 126398, https://doi.org/10.1016/j.biortech.2021.126398

Chen Q., Hills C. D., Tyrer M. and Slipper I. J. (2009). Immobilisation of heavy metal in cement-based solidification/stabilisation: a review. *Waste Management*, **29**(1), 390–403, https://doi.org/10.1016/j.wasman.2008.01.019

Darries G., Ayeleso A. and Raji A. (2022). Exploring hydropower options at a wastewater treatment plant – a case study. In Proceedings of the 30th Southern African Universities Power Engineering Conference (SAUPEC), Durban, South Africa, 1–6. IEEE. https://doi.org/10.1109/SAUPEC55179.2022.9730667

Delgado A., Foster S., Roman D. and Pahl-Wostl C. (2024). Water in circular economy and resilience (WICER) framework. *Utilities Policy*, **87**, 101604, https://doi.org/10.1016/j.jup.2023.101604

Ebrahimian F., Karimi K. and Kumar R. (2020). Sustainable biofuels and bioplastic production from the organic fraction of municipal solid waste. *Waste Management*, **116**, 40–48, https://doi.org/10.1016/j.wasman.2020.07.049

Finnveden G., Hauschild M. Z., Ekvall T., Guinée J., Heijungs R., Hellweg S., Koehler A., Pennington D. and Suh S. (2009). Recent developments in life cycle assessment. *Journal of Environmental Management*, **91**(1), 1–21, https://doi.org/10.1016/j.jenvman.2009.06.018

Gao H., Scherson Y. D. and Wells G. F. (2014). Towards energy neutral wastewater treatment: methodology and state of the art. *Environmental Science: Processes & Impacts*, **16**(6), 1223–1246, https://doi.org/10.1039/C4EM00069B

Gebrezgabher S., Rao K., Hanjra M. A. and Hernández-Sancho F. (2015). Business models and economic approaches for recovering energy from wastewater and fecal sludge. In Wastewater: Economic asset in an urbanizing world, P. Drechsel, M. Qadir and D. Wichelns (eds), Springer, Dordrecht, 215–232. https://doi.org/10.1007/978-94-017-9545-6

Gong A., Wang G., Qi X., He Y., Yang X., Huang X. and Liang P. (2025). Energy recovery and saving in municipal wastewater treatment engineering practices. *Nature Sustainability*, **8**, 112–119, https://doi.org/10.1038/s41893-024-01478-5

Govindarajan V. and Elmi R. A. (2013). Economic-environmental analysis of handling biogas from sewage sludge digesters in wastewater treatment plants for energy recovery: case study of Bekkelaget wastewater treatment plant in Oslo, Norway. *Energy*, **58**, 220–235, https://doi.org/10.1016/j.energy.2013.05.028

Jalil A. F. A., Pasupuleti J. and Khan M. R. B. (2019). Feasibility evaluation of micro-hydropower system generation for different types of sewage treatment plants. *International Journal of Recent Technology and Engineering*, **8**(3), 6338–6341.

Jiang X., Akbar A., Hysa E. and Akbar M. (2023). Environmental protection investment and enterprise innovation: Evidence from Chinese listed companies. *Kybernetes*, **52**, 708–727. https://doi.org/10.1108/K-12-2021-1292

Lassner K. J. (2011). Financial and Economic Analyses of Biogas-to-Energy Projects in Brazil. Duke University, Durham.

Li Y. and Williams P. T. (2025). Catalytic conversion of biomass components and waste biomass for hydrogen/syngas production using biochar catalysts. *Biomass and Bioenergy*, **194**, 107675. https://doi.org/10.1016/j.biombioe.2025.107675

Lima D., Li L. and Appleby G. (2024). A review of renewable energy technologies in municipal wastewater treatment plants (WWTPs). *Energies*, **17**(5), 6084, https://doi.org/10.3390/en17236084

Liu J., Chen Q. and Qi P. (2020). Upgrading of biogas to methane based on adsorption. *Processes*, **8**(8), 941, https://doi.org/10.3390/pr8080941

Liu L. X., Clegg S. and Pollack J. (2023). The effect of public–private partnerships on innovation in infrastructure delivery. *Project Management Journal*, **55**, 1–19. https://doi.org/10.1177/87569728231189989

Malik S., Singh R., Kumar A., Shukla A. and Yadav P. (2023). A perspective review on microbial fuel cells in treatment and product recovery from wastewater. *Water*, **15**(2), 316, https://doi.org/10.3390/w15020316

Mara D. (2013). Domestic Wastewater Treatment in Developing Countries. Routledge, London, UK. https://doi.org/10.4324/9781849771023

Mateo-Sagasta J., Raschid-Sally L. and Thebo A. (2015). Global wastewater and sludge production, treatment and use. In Wastewater: Economic asset in an urbanizing world, P. Drechsel, M. Qadir and D. Wichelns (eds), pp. 15–38. Springer Netherlands.

McCarty P. L., Bae J. and Kim J. (2011). Domestic wastewater treatment as a net energy producer–Can this be achieved? *Environmental Science & Technology*, **45**(17), 7100–7106, https://doi.org/10.1021/es2014264

Meena R. A. A., Kannah R. Y., Sindhu J., Ragavi J., Kumar G., Gunasekaran M. and Banu J. R. (2019). Trends and resource recovery in biological wastewater treatment system. *Bioresource Technology Reports*, **7**, 100235, https://doi.org/10.1016/j.biteb.2019.100235

Mekonnen M. M. and Hoekstra A. Y. (2016). Four billion people facing severe water scarcity. *Science Advances*, **2**, e1500323, https://doi.org/10.1126/sciadv.1500323

Musie W. and Gonfa G. (2023). Fresh water resource, scarcity, water salinity challenges and possible remedies: a review. *Heliyon*, **9**, e19031.

Nagpal H., Dev P., Sharma M. and Singh R. (2021). Heat recovery from wastewater–A review of available resource. *Water*, **13**(5), 1274, https://doi.org/10.3390/w13091274

Nizami A. S., Rehan M., Waqas M., Naqvi M., Ouda O. K. M., Shahzad K., Miandad R., Khan M. Z., Syamsiro M., Ismail I. M. I. and Pant D. (2017). Waste biorefineries: Enabling circular economies in developing countries. *Bioresource Technology*, **241**, 1101–1117. https://doi.org/10.1016/j.biortech.2017.05.097

Okadera T., Syutsubo K., Yoochatchaval W., Ebie Y. and Kubota R. (2020). Water volume- and BOD-based flow analysis for domestic wastewater treatment using wastewater inventories of Bangkok, Thailand. *Journal of Water and Environment Technology*, **18**, 71–79, https://doi.org/10.2965/jwet.19-064

Oladejo J., Shi K., Luo X., Yang G. and Wu T. (2018). A review of sludge-to-energy recovery methods. *Energies*, **12**, 60, https://doi.org/10.3390/en12010060

Pant D., Singh A., Van Bogaert G., Gallego Y. A., Diels L. and Vanbroekhoven K. (2011). An introduction to the life cycle assessment (LCA) of bioelectrochemical systems (BES) for sustainable energy and product generation: relevance and key aspects. *Renewable and Sustainable Energy Reviews*, **15**, 1305–1313. https://doi.org/10.1016/j.rser.2010.10.005

Piergrossi V., Canziani R., Capodaglio A. G. and Vaccari M. (2018). Energy recovery capacity evaluation within innovative biological wastewater treatment process. *Energy Conversion and Management*, **172**, 529–539, https://doi.org/10.1016/j.enconman.2018.07.027

Power C., McNabola A. and Coughlan P. (2014). Development of an evaluation method for hydropower energy recovery in wastewater treatment plants: case studies in Ireland and the UK. *Sustainable Energy Technologies and Assessments*, **7**, 166–177, https://doi.org/10.1016/j.seta.2014.06.001

Sangamnere R., Misra T., Bherwani H., Kapley A. and Kumar R. (2023). A critical review of conventional and emerging wastewater treatment technologies. *Sustainable Water Resources Management*, **9**, 58, https://doi.org/10.1007/s40899-023-00829-y

Sarpong G. and Gude V. G. (2021). Energy consumption and recovery in wastewater treatment systems. In: Resource Recovery From Wastewater, V. G. Gude (ed.), Apple Academic Press, New York, pp. 1–30.

Sikombe S., Muleya F., Phiri J., Zulu S., Simasiku P. and Situtu M. (2024). Key elements for promoting public-private partnerships in research and innovation. *Cogent Business & Management*, **11**, 2401627. https://doi.org/10.1080/23311975.2024.2401627

Sinagra M., Galelli S., Piccardo A., Cipollina A. and Micale G. (2022). Low-head hydropower for energy recovery in wastewater systems. *Water*, **14**(11), 1649, https://doi.org/10.3390/w14101649

Singh P. and Kansal A. (2018). Energy and GHG accounting for wastewater infrastructure. *Resources, Conservation and Recycling*, **128**, 499–507, https://doi.org/10.1016/j.resconrec.2016.07.014

Smith V. H. and Schindler D. W. (2009). Eutrophication science: where do we go from here? *Trends in Ecology & Evolution*, **24**, 201–207, https://doi.org/10.1016/j.tree.2008.11.009

Sozen S. and Shaw I. (2002). The international applicability of 'new' public management: lessons from Turkey. *International Journal of Public Sector Management*, **15**, 475–486. https://doi.org/10.1108/09513550210439625

Stillwell A. S., Hoppock D. C. and Webber M. E. (2010). Energy recovery from wastewater treatment plants in the United States: a case study of the energy-water nexus. *Sustainability*, **2**(4), 945–962, https://doi.org/10.3390/su2040945

Tang J., Chen Y., Zheng X. and Zhou Q. (2019). Municipal wastewater treatment plants coupled with electrochemical, biological and bio-electrochemical technologies: opportunities and challenge toward energy self-sufficiency. *Journal of Environmental Management*, **234**, 396–403, https://doi.org/10.1016/j.jenvman.2018.12.097

Tsalas N., Alexakis D., Vantarakis A. and Mamais D. (2024). Optimization of energy consumption in a wastewater treatment plant: an overview. *Energies*, **17**(5), 2808, https://doi.org/10.3390/en17122808

Tonini D. and Astrup T. (2012). LCA of biomass-based energy systems: a case study for Denmark. *Applied Energy*, **99**, 234–246. https://doi.org/10.1016/j.apenergy.2012.03.006

Twi-Yeboah N., Yusif I., Asare R., Oteng D. and Frimpong E. (2024). Enhancing energy efficiency and resource recovery in wastewater treatment plants. *Energies*, **17**(5), 3060, https://doi.org/10.3390/en17133060

Walker N. L., Williams A. P. and Styles D. (2021). Pitfalls in international benchmarking of energy intensity across wastewater treatment utilities. *Journal of Environmental Management*, **300**, 113613, https://doi.org/10.1016/j.jenvman.2021.113613

Zaman A. U. and Lehmann S. (2013). The zero waste index: a performance measurement tool for waste management systems in a 'zero waste city'. *Journal of Cleaner Production*, **50**, 123–132. https://doi.org/10.1016/j.jclepro.2012.11.041

Zhang Z., Li X., Yang Y., Zhao Y. and Chen G. (2022). Contaminant removal and resource recovery in bioelectrochemical wastewater treatment. *Current Pollution Reports*, **8**(2), 159–176, https://doi.org/10.1007/s40726-022-00218-7

doi: 10.2166/9781789065404_0111

Chapter 6

Data analytics, digitalization, and decision-support systems in wastewater management

*Wisdom Archford Munzeiwa[1], Turgay Cetinkaya[2], Suhasini Jindal[3], Harshita Jain[3] and Lovepreet Singh[4]**

[1]Chemistry Department, Bindura University of Science Education, Private Bag 1020, Bindura, Zimbabwe. E-mail: wmunzeiwa@buse.ac.zw
[2]Department of Aquatic Biotechnology and Genomics, Faculty of Aquatic Sciences, Istanbul University, Istanbul 34134, Türkiye. E-mail: turgaycetinkaya@istanbul.edu.tr
[3]Department of Climate Change and Environmental Sustainability, Amity Institute of Environment and Sustainability, Amity University Uttar Pradesh (AUUP), Noida, Uttar Pradesh 201301, India. E-mail: hjain@amity.edu
[4]Department of Chemical Engineering, Thapar Institute of Engineering and Technology, Patiala 147004, India. E-mail: lovepreet.singh@thapar.edu
**Corresponding author: lovepreet.singh@thapar.edu*

ABSTRACT

Recent advances in wastewater treatment systems have made data an integral part of planning, operation, and long-term management. The traditional methods based on periodic sampling and reactive decision-making are no longer able to cope with changing influent loads, energy constraints, and tightening regulations. This chapter reports on how data analytics and digitalization are changing wastewater management in ways that enable more informed, timely, and adaptive decisions. This chapter also examines how real-time data acquisition, advanced analytics, and decision-support systems can help improve process efficiency and operation reliability. The emphasis is on the application of predictive models, performance indicators and digital platforms to translate complex data into actionable insights for operators and planners. Instead of presenting digital tools as part of a solution, the chapter reflects how useful they are when incorporated into existing infrastructure and human expertise. In addition, concern is raised with data quality, system interoperability, and institution readiness, particularly with limited resources. This chapter draws out the potential for technology as well as the actuals of implementation by emphasizing how data-driven approaches can lead to sustainable, resilient, and transparent wastewater management systems.

Keywords: Data analytics, digitalization, decision-support systems, predictive modeling, smart wastewater management, system optimization

6.1 INTRODUCTION

Wastewater management today is an increasingly uncertain and complex process. With rapid urbanization, industrial diversification, population growth, and changes in water-use patterns, treatment plants must deal with increasing amounts of wastewater that are highly variable in characteristics (Beni *et al.*, 2025). Climate change has created an additional layer of challenge, where intense rainfall events cause sudden hydraulic overloads and prolonged dry periods lead to the concentration of pollutant loads (Ma *et al.*, 2020). Along with these operational requirements, wastewater utilities must also comply with increasingly stringent discharge standards, manage odors and emissions, limit energy consumption, and demonstrate transparency and accountability to regulators and the public (Perfect Pollucon Services, 2025). Under such circumstances, traditional management strategies based primarily on periodic sampling, routine inspections, and standardized operating procedures are becoming insufficient (García *et al.*, 2024). These approaches often capture only a snapshot of system performance and fail to reflect long-term fluctuations that may significantly influence treatment efficiency (Li *et al.*, 2024). As a result, operational decisions tend to be reactive, with interventions occurring only after process disturbances, compliance failures, or equipment malfunctions have already taken place. This reactive approach increases operational costs and energy consumption and may compromise effluent quality and system reliability (Hussain *et al.*, 2024).

Information technology has also played a key role in addressing these limitations. Online sensors, computerized monitoring systems, and digital control platforms provide a continuous stream of data, enabling real-time observation of wastewater treatment processes (Forhad *et al.*, 2024). Parameters such as flow rate, dissolved oxygen concentration, nutrient levels, energy consumption, and equipment status can be continuously tracked, allowing operators to better understand system behavior and anticipate when minor deviations may escalate into major operational problems (Newhart *et al.*, 2019). This transition from retrospective assessment to real-time awareness represents a significant advancement in wastewater system management (Ntalaperas *et al.*, 2022). The value of such data extends beyond day-to-day operations. When systematically aggregated and maintained, operational datasets can reveal long-term performance trends, seasonal variations, and the overall impacts of operational strategies (Mohammadi *et al.*, 2024). These insights support more effective planning for maintenance scheduling, capacity expansion, and process optimization. Furthermore, they enable wastewater utilities to transition toward performance-based management, where operational decisions are guided by empirical evidence rather than assumptions or historical practices (Prabu *et al.*, 2025) (Table 6.1).

Importantly, the growing role of data does not diminish the importance of human expertise. Experienced operators and engineers remain essential for interpreting data, understanding local system behavior, and making context-sensitive decisions (Ntalaperas *et al.*, 2022). Data serves as a decision-support tool that complements professional judgment rather than replacing it (Newhart *et al.*, 2019). When operators are equipped with reliable and timely information, their ability to manage complex wastewater treatment systems

Table 6.1 Types of operational data and their role in wastewater management.

Data Type	Source/Sensor	Typical Measurement Frequency	Management Relevance
Flow rate	Flow meters in influent/effluent lines	Continuous/ 5–15 min intervals	Detect hydraulic overloads, optimize pump scheduling, and maintain compliance with discharge limits
pH	Inline pH sensors	Continuous	Ensure chemical dosing is correct, maintain biological treatment efficiency, and prevent corrosion
Dissolved oxygen (DO)	DO probes in aeration tanks	Continuous	Optimize aeration energy use, maintain biological treatment efficiency, prevent oxygen deficits
Turbidity/TSS	Turbidity sensors, total suspended solids (TSS) meters	Continuous/ hourly	Track solids removal efficiency, prevent clogging, monitor effluent quality
Nutrients (N, P)	Online nitrate, ammonium, and phosphate analyzers	15–60 min intervals	Support nutrient removal optimization, regulatory compliance, and predictive modeling
Temperature	Thermistors or digital temperature sensors	Continuous	Biological reaction rates depend on temperature; guides process control and energy optimization
Energy consumption	Smart meters on pumps, aerators, and blowers	Continuous/ 15 min intervals	Monitor operational costs, identify inefficiencies, support energy-saving strategies
Equipment status/Alarms	SCADA system sensors, vibration or current sensors on motors	Real-time	Early fault detection, preventive maintenance scheduling, reduce downtime
Chemical dosing	Flow-proportional dosing sensors	Real-time/batch	Optimize reagent use, control treatment costs, ensure compliance
Influent characteristics (BOD, COD, SS)	Online analyzers or lab sampling	2–4 h intervals (or daily)	Adjust treatment process, predict operational loads, support predictive models

improves significantly (Benettin *et al.*, 2023). In contemporary wastewater management, data has therefore shifted from being a secondary record-keeping function to a critical operational resource (Forhad *et al.*, 2024). Continuous data streams support more adaptive, efficient, and resilient systems by enabling

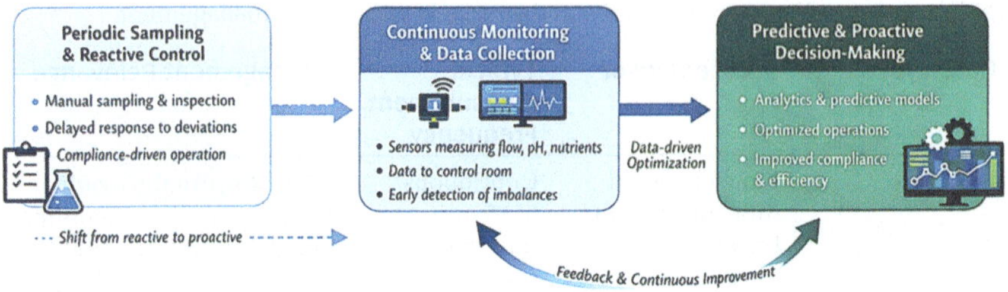

Figure 6.1 Transition from reactive to data-driven wastewater management.

informed decision-making at both operational and strategic levels (Prabu *et al.*, 2025). As wastewater systems continue to face increasing operational demands and environmental constraints, the effective use of data will be essential for ensuring environmental protection, regulatory compliance, and long-term sustainability (Newhart *et al.*, 2019) (Figure 6.1).

6.2 DIGITALIZATION OF WASTEWATER INFRASTRUCTURE

The digitalization of wastewater systems, which have traditionally been mechanical and manually operated, has become one of the primary drivers in the modernization of wastewater infrastructure. At the core of this transformation is the installation of sensors and monitoring devices that continuously measure critical parameters such as flow, pH, dissolved oxygen, turbidity, nutrient concentrations, energy consumption, and equipment performance. These sensors generate a continuous stream of information reflecting real-time operating conditions, allowing utilities to move beyond assumptions and respond to actual system behavior. Monitoring systems connect individual sensors across treatment plants, pumping stations, sewer networks, and receiving environments (Campisano *et al.*, 2002). Through wired or wireless communication systems, data from multiple assets can be transmitted to centralized control rooms or cloud-based platforms. This connectivity enables operators to manage large and complex wastewater systems through a unified interface, improving situational awareness and operational coordination. For utilities managing extensive sewer networks or multiple treatment facilities, such integrated monitoring systems are particularly valuable for identifying leaks, blockages, and abnormal flow patterns at an early stage before they result in service disruptions or environmental damage (Mohammadi *et al.*, 2024) (Figure 6.2).

Digital infrastructure generates large volumes of information, making data-acquisition systems essential for organizing and archiving this data effectively (Taleizadeh & Jalali Naini, 2018). Supervisory Control and Data Acquisition (SCADA) systems and related technologies collect raw sensor data, store it securely, and present it in formats that facilitate informed decision-making (Venkatasubramanian, 2019). These systems enable operators to visualize performance trends, compare the efficiency of different treatment

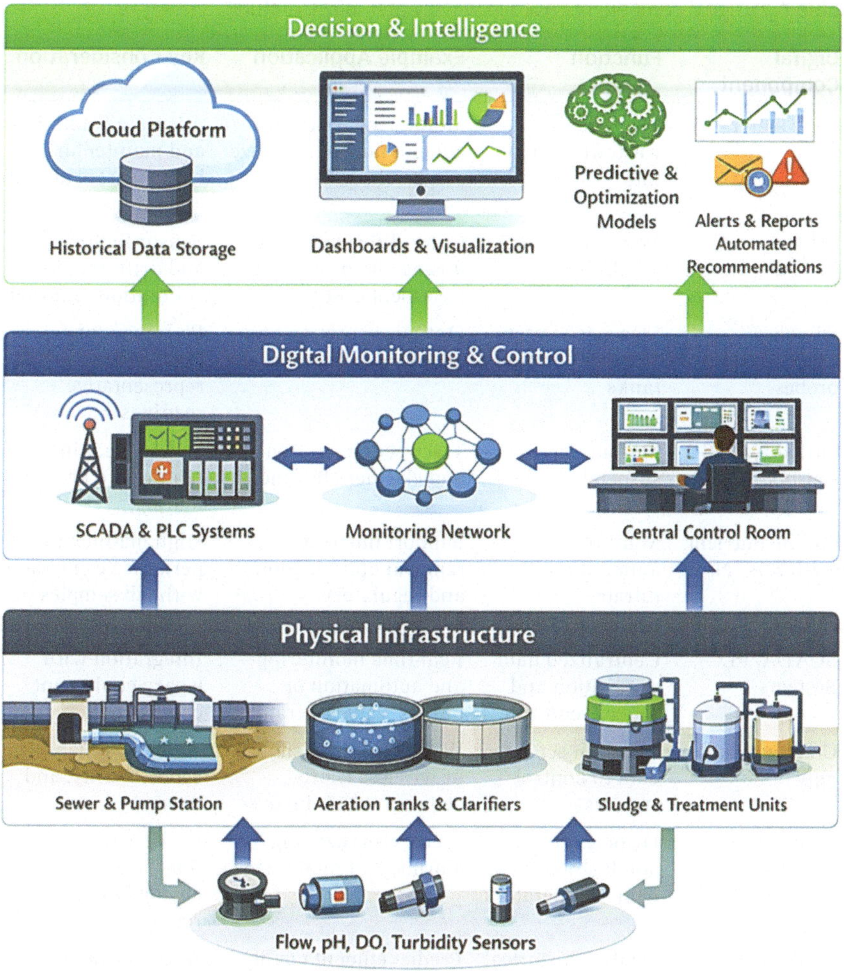

Figure 6.2 Integrated digital wastewater infrastructure.

units, and receive alerts when monitored parameters exceed predefined thresholds (Mohammadi *et al.*, 2024). Historical data retrieved from these platforms can further be used to evaluate operational performance, support regulatory reporting, and guide long-term process optimization (Stojanović & Milojković, 2016).

Digitalization is particularly transformative because it integrates physical infrastructure with intelligent digital tools. Advanced controllers embedded in pumps, valves, aerators, and other mechanical components increasingly adjust operational parameters based on real-time data. For instance, aeration rates can be dynamically optimized according to changing oxygen demand, thereby

Table 6.2 Digital components used in wastewater infrastructure.

Digital Component	Function	Example Application	Key Consideration
Flow meters	Measure wastewater flow rates	Monitoring influent/ effluent flow to optimize pump schedules	Accurate calibration and maintenance in harsh wastewater conditions
pH sensors	Monitor acidity/ alkalinity	Controlling chemical dosing and protecting treatment units	Sensor fouling and drift; regular calibration required
Dissolved oxygen (DO) probes	Measure oxygen levels in aeration tanks	Optimize aeration and energy consumption	Placement in tank critical for representative readings
Turbidity/TSS sensors	Measure suspended solids	Track sedimentation and clarifier efficiency	Sensor cleaning required to prevent fouling
Online nutrient analyzers (N, P)	Monitor ammonium, nitrate, phosphate	Support nutrient removal optimization and regulatory compliance	High maintenance; periodic verification with lab samples
SCADA/PLC Systems	Centralized data acquisition and process control	Real-time monitoring and automation of pumps, valves, aerators	Integration with legacy equipment can be challenging
Communication networks	Transmit sensor data to control systems	Wired (Ethernet, fiber) or wireless (LoRa, cellular, Wi-Fi) links	Network reliability, cybersecurity, and latency
Dashboards/ visualization platforms	Display real-time and historical data for operators	Trend visualization, alarms, KPI monitoring	User-friendly interface and customization required
Predictive modeling platforms	Analyze historical data to forecast performance	Predict effluent quality, optimize energy use, detect anomalies	Accuracy depends on quality and completeness of data
Cloud storage/ IoT platforms	Store and manage large datasets, enable remote access	Centralized data storage, remote monitoring, and analytics	Data security, privacy, and connectivity constraints

reducing energy consumption while maintaining treatment efficiency. This integration of physical and digital components enables more precise process control and minimizes the need for continuous manual intervention (Forhad *et al.*, 2024) (Table 6.2).

Despite its advantages, digitalization is not merely a matter of installing new technologies. Successful implementation requires careful consideration of system compatibility, maintenance requirements, and staff capacity (Ntalaperas *et al.*, 2022). Sensors must be reliable and robust enough to function effectively in harsh wastewater environments, data transmission systems must ensure security and integrity, and operators must be adequately trained to trust and utilize digital tools in their daily operations. When these elements are addressed in an integrated manner, digitalization strengthens the connection between physical infrastructure and information systems, enabling wastewater treatment facilities to operate more efficiently, transparently, and adaptively in response to future environmental and operational challenges (Forhad *et al.*, 2024).

6.3 ANALYTICAL TOOLS AND PREDICTIVE MODELING APPROACHES

Analytical tools play a critical role in transforming raw wastewater process data into meaningful knowledge. Data by itself has limited value unless it is thoroughly analyzed to uncover patterns, relationships, and system behavior (Singh & Sharma, 2023). A wide range of analytical approaches, from traditional statistical techniques to advanced machine learning models, are used to evaluate system performance, identify inefficiencies, and support informed operational decisions (Li *et al.*, 2025). Conventional statistical methods continue to be widely applied in wastewater treatment because they are relatively simple to implement and easy to interpret. Techniques such as trend analysis, correlation studies, regression modeling, and control charts help operators understand how key operational parameters respond to variations in influent characteristics or operational adjustments (Mohammadi *et al.*, 2024). For instance, statistical analyses can be used to estimate seasonal fluctuations in nutrient loads, assess regulatory compliance patterns, or evaluate the effects of process modifications on effluent quality. These approaches are particularly valuable for establishing baseline performance and supporting evidence-based decision-making in wastewater treatment operations (Rossi *et al.*, 2021) (Figure 6.3).

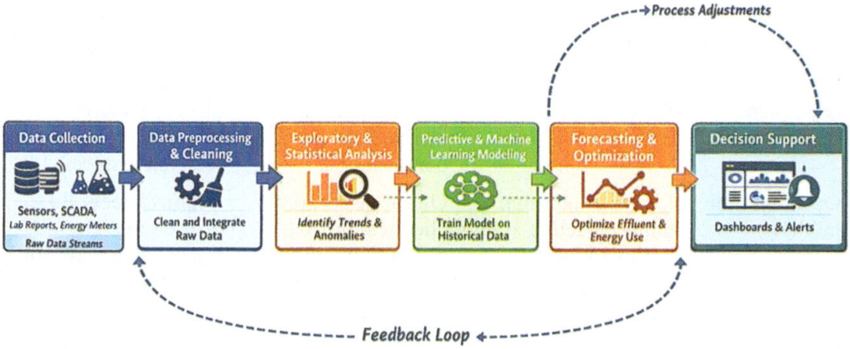

Figure 6.3 Workflow for data analytics and predictive modeling.

Recently, machine learning and data-driven modeling approaches have gained considerable attention for their ability to capture the complex and non-linear relationships inherent in biological and physicochemical wastewater treatment processes. Historical operational datasets can be used to develop predictive models such as artificial neural networks, decision trees, support vector machines, and ensemble learning methods, which are capable of forecasting system behavior under varying operational conditions. In practical applications, these models are increasingly used to predict effluent quality, anticipate equipment failures, and optimize energy-intensive operations such as aeration and pumping (Prabu *et al.*, 2025). One of the major advantages of these approaches is their scalability, as model performance and predictive accuracy tend to improve as larger datasets become available.

Table 6.3 Analytical and predictive methods applied in wastewater systems.

Method Category	Example Techniques	Application Area	Practical Limitation
Statistical analysis	Regression analysis, correlation studies, control charts, ANOVA	Trend detection, process performance evaluation, baseline assessment	Limited ability to capture non-linear or complex system behavior
Time-series forecasting	ARIMA, exponential smoothing	Predict influent flow/quality, effluent characteristics	Assumes stationary data; sensitive to missing values
Artificial neural networks (ANNs)	Feedforward ANN, LSTM	Predict effluent quality, process efficiency, energy demand	Requires large datasets; less interpretable; may overfit
Decision Trees & Random Forests	CART, random Forest, Gradient Boosted Trees	Predict system behavior, detect anomalies, optimize process settings	Can be computationally intensive; sensitive to noisy data
Support vector machines (SVM)	Regression SVM, Classification SVM	Fault detection, effluent quality classification	Requires careful parameter tuning; less intuitive for operators
Clustering & pattern recognition	K-Means, DBSCAN, PCA	Identify influent patterns, group similar process conditions	Does not provide direct prediction; results depend on selected features
Optimization models	Linear programming, multi-objective optimization	Optimize aeration, chemical dosing, pump scheduling	Requires accurate process models; may not capture real-time variability
Hybrid/ ensemble models	ANN + ARIMA, Random Forest + optimization	Improve prediction accuracy and robustness	Increased complexity; harder to implement in real-time operations

In addition to predictive modeling, forecasting tools further enhance analytical capabilities by enabling forward-looking operational planning. Short-term forecasts can support daily operational decisions, such as adjusting process controls in anticipation of peak influent loads or wet-weather flow events. Long-term forecasting, on the other hand, assists in strategic planning related to capacity expansion, infrastructure upgrades, and resource allocation (Newhart *et al.*, 2019). When combined with scenario analysis, these analytical frameworks also allow utilities to evaluate the potential impacts of alternative operational strategies or external pressures such as population growth and climate variability (Singh & Sharma, 2023) (Table 6.3).

Despite their potential, analytical and predictive tools must be applied carefully in wastewater management systems. The reliability of these models largely depends on the quality and completeness of the data used for their development, and overly complex analytical tools may become difficult to interpret, implement, or maintain in practical operational settings (Newhart *et al.*, 2019). Therefore, the successful application of data analytics requires a balance between analytical sophistication and practical usability (Rossi *et al.*, 2021). When these tools are integrated with the experience of system operators and a strong understanding of local treatment dynamics, analytical models can serve as powerful instruments for improving process efficiency, reducing operational risks, and enhancing the overall performance of wastewater treatment and management systems (Newhart *et al.*, 2019).

6.4 DECISION-SUPPORT SYSTEMS FOR PLANNING AND OPERATIONS

The decision-support systems (DSS) act as the point of contact between the gathering, analysis, and actions of wastewater treatment. Sensors and analytical equipment produce huge amounts of data, but decision-support platforms must show this information in terms of operators, engineers, and managers can understand and apply easily. Their primary value is bringing the large volumes of data into tangible and actionable information that can be used to make informed decisions and improve decision-making. Dashboards are one of the most widely used components of DSS. They provide a graphic representation of performance measures such as flow rates, treatment efficiency, energy consumption, and compliance status. Dashboards provide users with easy access to trends, alerts, and comparisons that give them an opportunity to quickly check system performance and avoid anomalies. Users can search by data source instead of going through multiple sources, focusing on critical variables that need to be looked at: that is, increased response time and operational confidence (Figure 6.4).

Control platforms extend the functionality of data analytics by linking data insights directly to operational actions. When integrated with SCADA and other automation systems, these platforms enable operators to modify process settings, schedule equipment operations, or trigger alarms based on predefined rules or real-time system conditions (Venkatasubramanian, 2019).

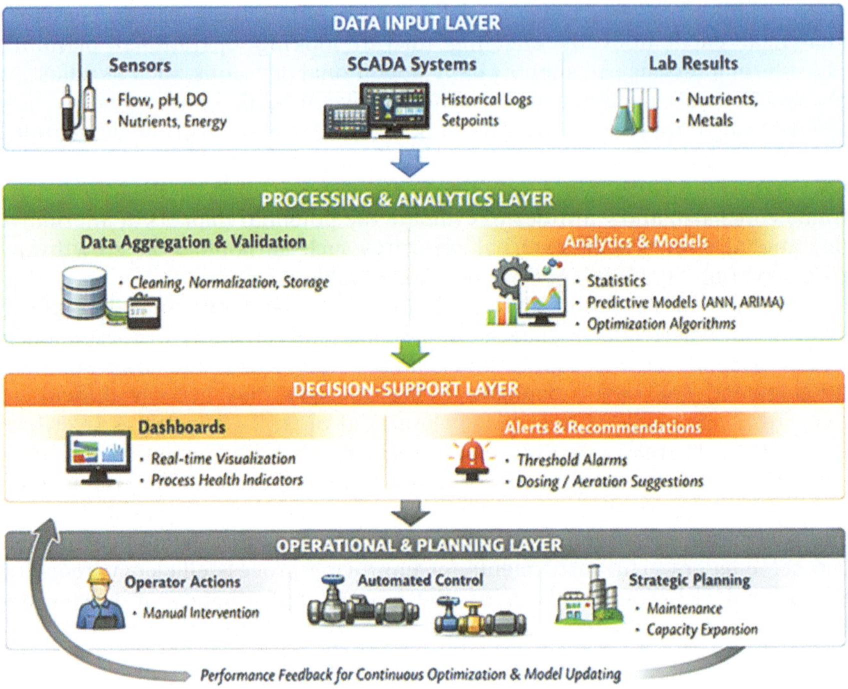

Figure 6.4 Decision-support system structure for wastewater management.

For instance, aeration intensity can be adjusted in response to fluctuations in oxygen demand, while pumping schedules can be optimized to reduce energy consumption during peak tariff periods. Such data-driven control strategies enhance process stability, reduce operational variability, and minimize the need for constant manual intervention.

Managerial decision-support tools also play an important role in assisting planners and decision-makers in evaluating long-term operational and investment strategies. These tools integrate historical datasets, predictive models, and scenario analysis to assess system performance under multiple operational or investment scenarios (World Bank, n.d.). They are particularly useful in areas such as capacity planning, asset management, and regulatory compliance, as utilities can justify their decisions through transparent and evidence-based assessments (World Bank, n.d.).

Importantly, DSS are not intended to replace human judgment but rather to enhance it. Their effectiveness depends on clear system design, reliable data inputs, and user-friendly interfaces that align with existing operational workflows (Kahneman & Klein, 2009). When implemented effectively, DSS enable wastewater professionals to make more informed, consistent, and proactive decisions, ultimately improving the efficiency, resilience, and sustainability of wastewater management systems (Table 6.4).

Table 6.4 Decision-support tools and their users.

Tool Type	Primary User	Decision Level	Key Output
Real-time dashboards	Operators, shift supervisors	Operational/ Tactical	Current process status, alerts for abnormal conditions, KPI visualization
SCADA/control platforms	Operators, process engineers	Operational	Automated control of pumps, valves, aeration, chemical dosing; real-time process adjustments
Predictive modeling platforms	Process engineers, planners	Tactical/ Strategic	Forecast influent flows, effluent quality, energy use; support proactive interventions
Maintenance management tools (CMMS)	Maintenance teams, managers	Operational/ Tactical	Equipment status, preventive maintenance schedules, fault detection reports
Performance analytics/KPI tools	Managers, planners	Tactical/ Strategic	Evaluate plant efficiency, regulatory compliance, energy performance, trend reports
Scenario analysis/ optimization tools	Planners, decision makers	Strategic	Simulate operational or investment scenarios, optimize resource allocation, plan expansions
Alerts/notification systems	Operators, supervisors	Operational	Early warning of critical deviations, alarms for process or equipment failures
Mobile monitoring apps/ remote access	Field staff, managers	Operational/ Tactical	Remote monitoring, instant notifications, data access on the go
Cloud-based integration platforms	IT teams, managers	Strategic	Centralized data management, integration of multiple plants, support for analytics and reporting

6.5 IMPLEMENTATION CHALLENGES AND INSTITUTIONAL CONSIDERATIONS

Data analytics and digital tools offer significant potential benefits, yet their effectiveness in wastewater treatment is often constrained by practical and institutional challenges (Ntalaperas *et al.*, 2022). One of the most fundamental issues is data quality. Sensors operating in harsh wastewater environments are prone to fouling, drift, and failure, which can lead to incomplete or unreliable datasets (Ntalaperas *et al.*, 2022). Confidence in digital outputs is further weakened by inconsistent calibration practices, data collection errors, and the absence of systematic validation procedures. Without reliable and high-quality data, even the most advanced analytical tools and DSS may produce misleading results.

Interoperability represents another significant concern, particularly in long-established wastewater systems. Many utilities operate a combination of legacy infrastructure and modern digital technologies that were not originally designed to communicate with one another (Singh & Sharma, 2023). Differences in data formats, software platforms, and communication protocols can make system integration costly and technically complex. As a result, data sharing across departments or facilities may be limited, restricting the potential for comprehensive system-wide analysis and coordinated decision-making (Kahneman & Klein, 2009; Singh & Sharma, 2023).

Digital adoption also depends strongly on human and institutional capacity (World Bank, 2016). The effective use of digital platforms requires personnel who possess the skills to operate, interpret, and maintain these technologies (Davis, 1989). However, many utilities–particularly in low- and middle-income regions–face constraints in training, technical expertise, and capacity-building programs (World Bank, 2020). Resistance to technological change may also arise due to concerns about system complexity, cybersecurity risks, or perceived threats to established roles and operational practices (OECD, n.d.). Addressing these challenges requires continuous training programs, clear communication strategies, and a gradual integration of digital tools into existing workflows (Kotter, 2007).

Socio-economic context also plays a critical role in shaping the adoption of digitalization strategies (Heeks, 2002). Utilities with limited financial and technical resources may struggle with the high upfront costs of digital infrastructure, limited technical support, and competing priorities such as expanding basic wastewater services (World Bank, n.d.). Conversely, well-resourced utilities may face challenges related to data governance, cybersecurity risks, and institutional coordination (European Union Agency for Cybersecurity (ENISA), n.d.). These variations highlight the need for context-specific implementation strategies rather than universal or one-size-fits-all solutions (Li *et al.*, 2023). Ultimately, the success of digital wastewater management depends as much on institutional readiness as on technological capability (Li *et al.*, 2023). Aligning digital initiatives with organizational objectives, regulatory frameworks, and local operational capacities is essential for effective implementation (OECD, n.d.). When these challenges are proactively addressed, digital tools can be integrated more successfully, supporting equitable, resilient, and sustainable wastewater management across diverse environmental and institutional contexts (International Water Association, n.d.).

6.6 CONCLUSION

The planning, operation, and management of wastewater systems is being reshaped through data analytics, digitalization, and decision support systems. As wastewater infrastructure is becoming more complex and work requirements continue to grow, reactive and experience-based approaches are no longer sufficient. Utilization of continuous data streams, analytical tools, and digital platforms provides utilities with a vantage point for understanding system behavior, to respond more adaptively to changing circumstances, and to develop longer-term sustainability plans. But it is not an automatic adoption of new technologies that unlocks the benefits of digital wastewater management. They have to be data-rich, dependable, and most importantly, useful to institutions and staff. Digital solutions are most useful when they complement the infrastructure and operational knowledge, rather than trying to replace them. The hybridity of technology and human experience allows the informational insights to be translated into practical and context-sensitive decisions. Looking ahead, the role of digitalization in wastewater management is likely to expand further, supporting greater transparency, resilience, and resource efficiency. By addressing implementation challenges and aligning digital strategies with local socio-economic realities, wastewater utilities can harness data not just as an operational asset, but as a foundation for sustainable and adaptive water management systems.

REFERENCES

Benettin G., Manfrida G. and Pilo F. (2023). Real-time monitoring and static data analysis to assess energetic and environmental performances in the wastewater sector: A case study. LAPSE: Linking Academic Publications to Sustainable Development, LAPSE-2023.17936. https://psecommunity.org/wp-content/plugins/wpor/includes/file/2303/LAPSE-2023.17936-1v1.pdf

Beni S., Farid M. and Aldeen M. (2025). Wastewater Management Challenges in Rapidly Growing Urban Areas. Netsol Water, India. https://www.sewagetreatmentplants.in/wastewater-management-challenges-in-rapidly-growing-urban-areas/

Campisano A., Creaco E. and Modica C. (2002). PID and PLC units for the real-time control of sewer systems. *Water Science and Technology*, **45**(7), 95–102. https://doi.org/10.2166/wst.2002.0172

Davis F. D. (1989). Perceived usefulness, perceived ease of use, and user acceptance of information technology. *MIS Quarterly*, **13**(3), 319–340. https://www.jstor.org/stable/249008

European Union Agency for Cybersecurity (ENISA). (n.d.). *Cybersecurity of critical sectors*. https://www.enisa.europa.eu/topics/cybersecurity-of-critical-sectors

Forhad H. M., Chowdhury M. M. H. and Rahman M. M. (2024). IoT-based real-time water quality monitoring system in water treatment plants (WTPs). *Environmental Monitoring and Assessment*, **196**(11), 1007. https://pmc.ncbi.nlm.nih.gov/articles/PMC11652906/

García J., Martínez F. and López A. (2024). A review of the state-of-the-art wastewater quality characterization and measurement technologies: Is the shift to real-time monitoring nowadays feasible? *Journal of Water Process Engineering*, **61**, 104–215. https://www.sciencedirect.com/science/article/pii/S2214714424002939

Heeks R. (2002). Information systems and developing countries: failure, success, and local improvisations. *The Information Society*, **18**(2), 101–112. https://doi.org/10.1080/01972240290075039; https://www.ppsthane.com/blog/cpcb-wastewater-discharge-standards

Hussain A., Khan S. and Ali R. (2024). Dynamic modelling, process control, and monitoring of selected biological and advanced oxidation processes for wastewater treatment: A review of recent developments. *Processes*, **12**(2), 345. https://pmc.ncbi.nlm.nih.gov/articles/PMC10886268/

International Water Association. (n.d.). Digital Water Programme. International Water Association. London, UK, https://www.iwa-network.org/projects/digital-water

Kahneman D. and Klein G. (2009). Conditions for intuitive expertise: A failure to disagree. *American Psychologist*, **64**(6), 515–526. https://doi.org/10.1037/a0016755

Kotter J. P. (2007). Leading change: Why transformation efforts fail. *Harvard Business Review*, **85**(1), 96–103. https://hbr.org/2007/01/leading-change-why-transformation-efforts-fail

Li R., Zhang Y. and Chen J. (2023). Smart water management: governance innovation, technological integration, and policy pathways toward economic and ecological sustainability. *Water*, **17**(13), 1932. https://doi.org/10.3390/w17131932

Li Y., Zhang H. and Chen G. (2024). A review of pollution-based real-time modelling and control for sewage systems. *Water Research*, **250**, 120123. https://pmc.ncbi.nlm.nih.gov/articles/PMC11214442/

Li Y., Zhang H. and Chen G. (2025). Machine learning in wastewater treatment: A comprehensive bibliometric review. *ACS ES&T Water*, **5**(2), 635–651. https://doi.org/10.1021/acsestwater.4c01047

Ma X., Wang Z., Xu H., Li Y. and Li F. (2020). Assessing environmental impacts of large centralized wastewater treatment plants with combined or separate sewer systems in dry/wet seasons by using LCA. *Environmental Science and Pollution Research*, **27**(15), 18310–18324. https://pmc.ncbi.nlm.nih.gov/articles/PMC7190677/

Mohammadi E., Hansen J. H. and Mikkelsen P. S. (2024). Wastewater treatment plant data for nutrient removal system. *arXiv Preprint*, arXiv:2407.05346. https://arxiv.org/abs/2407.05346

Newhart K. B., Holloway R. W., Hering A. S. and Cath T. Y. (2019). Data-driven performance analyses of wastewater treatment plants: A review. *Water Research*, **157**, 498–513. https://doi.org/10.1016/j.watres.2019.03.030

Ntalaperas D., Kaselimi M., Doulamis A., Vayas I. and Doulamis N. (2022). Intelligent tools to monitor, control and predict wastewater reclamation and reuse. *Applied Sciences*, **12**(8), 3870. https://doi.org/10.3390/app12083870

Organisation for Economic Co-operation and Development. (n.d.). Governance. OECD. Paris, France. https://www.oecd.org/en/topics/governance.html

Perfect Pollucon Services (2025, September 27). CPCB Wastewater Discharge Standards Explained. Perfect Pollucon Services. Majiwada Thane, Maharashtra, India.

Prabu P., Kumar S. and Gupta R. (2025). AquaFlowNet: A machine learning based framework for real-time wastewater flow management and optimization. *Scientific Reports*, **15**, 99200. https://doi.org/10.1038/s41598-025-99200-8

Rossi F., Boldo M. and Romano M. (2021). Data analytics and pump control in a wastewater treatment plant. *Applied Energy*, **295**, 117031. https://doi.org/10.1016/j.apenergy.2021.117031

Singh R. and Sharma P. (2023). Waste to wisdom: A review paper on dataflow–optimizing wastewater treatment with analytics. *International Journal of Research Publication and Reviews*, **4**(11), 1079–1086. https://ijrpr.com/uploads/V4ISSUE11/IJRPR19262.pdf

Stojanović B. and Milojković D. (2016). Analysis of insolation potential of Knjaževac municipality (Serbia) using multi-criteria approach. *Renewable and Sustainable Energy Reviews*, **63**, 103–115. https://doi.org/10.1016/j.rser.2016.05.030

Taleizadeh A. A. and Jalali Naini S. G. (2018). Reverse channel selection decisions with a joint third-party recycler. *International Journal of Production Research*, **56**(18), 6035–6057. https://doi.org/10.1080/00207543.2018.1442944

Venkatasubramanian V. (2019). Perspectives on process monitoring of industrial systems. *Computers & Chemical Engineering*, **126**, 288–300. https://doi.org/10.1016/j.compchemeng.2019.04.008

World Bank. (2016). World Development Report 2016: Digital Dividends. World Bank. Washington DC, USA. https://www.worldbank.org/en/publication/wdr2016

World Bank. (2020). Utility of the Future Program. World Bank. Washington DC, USA. https://www.worldbank.org/en/topic/water/publication/utility-of-the-future

World Bank. (n.d.). Development Topics. World Bank. https://www.worldbank.org/ext/en/development-topics

doi: 10.2166/9781789065404_0125

Chapter 7

Federated learning for secure and privacy-preserving edge AI in smart cities' wastewater management systems

Varad Joshi[1], Shadab Siddique[1], Sumit Hazra[1] and Shahin Fatima[1]*

Department of Computer Science and Engineering, Koneru Lakshmaiah Education Foundation, Hyderabad–500075, Telangana, India
**Corresponding author: varadj40@gmail.com*

ABSTRACT

The rapid evolution of smart cities has significantly transformed wastewater management systems through the integration of real-time monitoring, intelligent control, and data-driven decision-making. However, conventional centralized artificial intelligence (AI) frameworks face critical limitations, including data privacy vulnerabilities, security risks, high communication overhead, and poor scalability across geographically distributed treatment facilities. This study investigates federated learning (FL) as an enabling paradigm for developing secure, privacy-preserving, and scalable edge AI solutions tailored to smart wastewater infrastructure. Unlike traditional centralized approaches, FL facilitates decentralized model training across wastewater treatment plants by sharing model parameters instead of raw data, thereby reducing data transmission costs and ensuring compliance with data protection regulations. To enhance system robustness, a hybrid framework is proposed that integrates differential privacy, secure multi-party computation (SMPC), and blockchain-based authentication. These mechanisms collectively strengthen resilience against adversarial attacks, prevent data leakage, and ensure the integrity of model updates. Furthermore, an adaptive aggregation strategy is introduced to address data heterogeneity across distributed facilities by dynamically weighting model contributions based on sensor reliability, data quality, and network conditions. The proposed framework is evaluated using representative wastewater datasets, demonstrating significant improvements in anomaly detection accuracy, robustness to non-identically distributed data, and privacy preservation compared to conventional centralized AI models. Additionally, the framework achieves enhanced energy efficiency and reduced communication overhead in edge environments. The findings establish FL as a foundational technology for deploying secure, scalable, and privacy-aware AI systems in smart wastewater management.

Keywords: Federated learning, Edge AI, smart wastewater management, privacy preservation, anomaly detection, secure AI, blockchain

7.1 INTRODUCTION

The global transition toward smart cities has accelerated the integration of advanced digital technologies, including the Internet of Things (IoT), edge computing, and artificial intelligence (AI), into critical urban infrastructure. Wastewater management, as a fundamental pillar of sustainable urban systems, increasingly relies on intelligent monitoring and automated control mechanisms to ensure efficient resource utilization, regulatory compliance, and public health protection (Bibri *et al.*, 2023). Despite these advancements, conventional centralized AI frameworks exhibit significant limitations when deployed across large-scale and geographically distributed wastewater networks. These limitations include high communication overhead, vulnerability to single points of failure, and increased exposure to data privacy and cybersecurity threats (Albshaier, 2025). Such challenges restrict the scalability and robustness of centralized systems in real-world smart city environments.

Federated learning (FL) has emerged as a promising paradigm for decentralized model training, enabling data to remain localized at the source while only model parameters are shared across the network (Deng *et al.*, 2025). This approach mitigates privacy risks, reduces network congestion, and supports scalability in heterogeneous environments by eliminating the need for raw data transfer (Kalapaaking *et al.*, 2023). When integrated with edge AI, FL enables real-time analytics and rapid decision-making capabilities, which are essential for critical applications such as anomaly detection in wastewater streams and predictive system monitoring (Rane *et al.*, 2024). However, the implementation of FL in wastewater management systems is not without challenges. Data heterogeneity across treatment facilities, fluctuating network conditions, and the risk of adversarial attacks on model updates can adversely impact system performance and reliability (Nicolazzo, 2024) (Figure 7.1).

To address these challenges, recent research highlights the importance of integrating advanced security and privacy-preserving mechanisms within FL frameworks. Techniques such as differential privacy (DP), secure multi-party computation (SMPC), and blockchain-based authentication have

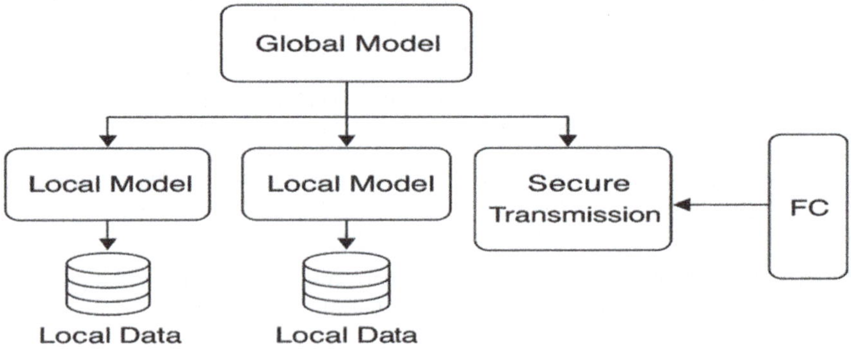

Figure 7.1 Federated learning architecture for smart wastewater management.

demonstrated strong potential in enhancing data confidentiality, preventing unauthorized access, and ensuring the integrity of model updates (Limbepe, 2025). Furthermore, adaptive aggregation strategies have been proposed to improve model performance in heterogeneous environments. These methods dynamically adjust the contribution of individual client models based on parameters such as data quality, sensor reliability, and network stability, thereby enhancing the robustness and generalization capability of the global model (Zhan *et al.*, 2025). In this study, a hybrid FL framework is proposed for smart wastewater management systems, integrating privacy-preserving techniques with adaptive aggregation mechanisms to address domain-specific operational and security challenges.

7.2 MATERIALS AND METHODS

7.2.1 System overview

The proposed system is designed as a hybrid FL architecture to enable secure, privacy-preserving, and efficient edge AI deployment for smart wastewater management systems. The framework integrates a distributed model training pipeline based on federated principles, ensuring that data remains localized within individual wastewater treatment facilities. In addition, a multi-layer privacy and security module is incorporated, combining DP, SMPC, and blockchain-based authentication to safeguard data integrity and confidentiality. To further enhance system performance in heterogeneous environments, an adaptive aggregation mechanism is introduced to account for variations in data quality, sensor reliability, and network conditions. This architecture is particularly suited to wastewater treatment systems, which are inherently decentralized and operate under diverse infrastructural and environmental conditions.

7.2.2 Federated learning architecture

7.2.2.1 Client–server model

The FL system follows a centralized coordination model with multiple distributed client nodes, where each wastewater treatment plant (WWTP) functions as an independent client equipped with edge computing capabilities. These client nodes perform local model training using their respective datasets, thereby eliminating the need to transfer raw data to a central repository. Instead, only encrypted model parameters or gradients are communicated to a central aggregation server. The server aggregates these updates to construct a global model, which is subsequently redistributed to all participating clients for further training iterations. This iterative process continues until the model converges to an optimal state. Such an architecture enables collaborative intelligence while ensuring compliance with data privacy regulations and minimizing the risks associated with centralized data storage (McMahan *et al.*, 2017).

7.2.2.2 Communication workflow

The communication workflow in the proposed FL system follows a structured sequence of operations within each training round. Initially, the central server

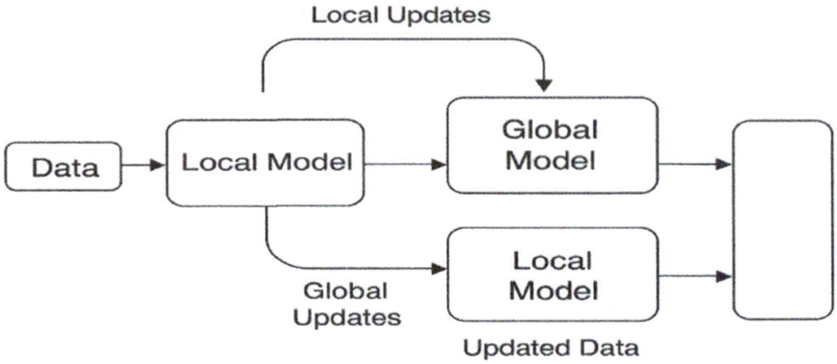

Figure 7.2 End-to-end training cycle.

distributes the current global model to all client nodes. Each wastewater facility then performs local training using its own dataset, capturing site-specific characteristics and operational patterns. Before transmission, local model updates undergo privacy-preserving transformations through DP and SMPC techniques. These transformed updates are then securely transmitted to the central server using encrypted communication channels. Upon receiving updates from all participating clients, the server performs adaptive aggregation based on predefined weighting criteria and generates an updated global model. This updated model is subsequently redistributed to all clients, completing one training cycle (Figure 7.2).

7.2.3 Privacy and security framework

7.2.3.1 Differential privacy

DP ensures that the inclusion or exclusion of a single data point does not significantly affect the output of the model. It is mathematically defined as

$$Pr[M(D) \in S] \leq e^{\epsilon} \cdot Pr[M(D') \in S] + \delta \tag{7.1}$$

where M represents the randomized mechanism, D and D' are adjacent datasets differing by one record, ϵ is the privacy budget, and δ denotes the probability of privacy guarantee failure (Dwork & Roth, 2014). In this study, an adaptive privacy budget strategy is employed, where lower noise is applied during initial training rounds to stabilize learning, while stronger privacy constraints are enforced in later stages to enhance data protection.

7.2.3.2 Secure multi-party computation

SMPC enables collaborative aggregation without revealing individual model updates. Using additive secret sharing, each client divides its update W_i into multiple shares:

$$W_i = \sum_{k=1}^{K} s_{i,k} \tag{7.2}$$

where $s_{i,k}$ represents the k-th share distributed among participating nodes. The server only receives the aggregated sum:

$$\sum_{i=1}^{N} W_i = \sum_{i=1}^{N}\sum_{k=1}^{K} s_{i,k} \tag{7.3}$$

This ensures that no individual update can be reconstructed without full collusion among participants (Yao, 1982).

7.2.3.3 Blockchain-based authentication

To ensure the integrity and traceability of model updates, a lightweight blockchain-based authentication layer is integrated into the FL pipeline. Each model update is cryptographically signed and recorded in a distributed ledger, which maintains an immutable record of all transactions. The blockchain stores essential metadata, including timestamps, hashed summaries of model updates, and digital signatures of participating entities. A proof-of-authority consensus mechanism is employed to achieve low-latency validation, making it suitable for real-time municipal infrastructure applications. This approach effectively prevents model poisoning attacks and ensures that all updates can be audited and verified when necessary (Li *et al.*, 2021).

7.2.4 Adaptive aggregation mechanism

7.2.4.1 Rationale

In wastewater management systems, significant heterogeneity exists across treatment facilities in terms of sensor accuracy, operational conditions, and data quality. Traditional aggregation methods such as federated averaging treat all client updates equally, which can negatively impact model performance when some datasets are noisy or biased. To address this limitation, the proposed framework incorporates an adaptive aggregation mechanism that assigns dynamic weights to each client's contribution based on contextual parameters.

7.2.4.2 Aggregation strategy

The global model update is computed using a weighted aggregation approach to address heterogeneity across WWTPs. The aggregation is defined as

$$W_{\text{global}} = \sum_{i=1}^{N} \omega_i W_i \tag{7.4}$$

where W_i is the local model update from client i, and ω_i represents the adaptive weight assigned to that client.

The weight factor is calculated as

$$\omega_i = \frac{\alpha_i \cdot \beta_i \cdot \gamma_i}{\sum_{j=1}^{N}(\alpha_j \cdot \beta_j \cdot \gamma_j)} \tag{7.5}$$

where:
α_i denotes the sensor reliability score (SRS),
β_i represents the data quality index (DQI), and
γ_i corresponds to the network stability factor (NSF).

This formulation ensures that clients with higher data quality and reliable communication contribute more significantly to the global model.

7.2.5 Dataset and preprocessing

7.2.5.1 Data sources

The proposed framework is validated using datasets collected from multiple municipal WWTPs. These datasets include time-series sensor data such as flow rate, pH, temperature, and chemical concentrations, along with historical maintenance logs and energy consumption records. Such diverse data sources enable comprehensive analysis of both operational and environmental parameters within wastewater systems.

7.2.5.2 Data preprocessing

Data preprocessing is performed to ensure the quality and consistency of input data for model training. Missing values are addressed using interpolation techniques for short gaps and K-nearest-neighbor (KNN) imputation for more complex missing patterns. Outliers are identified and removed based on interquartile range analysis and domain-specific thresholds. Feature scaling is performed using Min–Max normalization and Z-score standardization to ensure compatibility with neural network models. Additionally, noise in sensor readings is reduced using Fourier filtering techniques, improving signal clarity for downstream analysis.

7.2.6 Model architecture

The global model update during each training iteration follows a gradient-based optimization process:

$$W^{t+1} = W^t - \eta \sum_{i=1}^{N} \omega_i \nabla L_i(W^t) \tag{7.6}$$

where W^t represents the model parameters at iteration t, η is the learning rate, ω_i is the adaptive weight factor, and $\nabla L_i(W^t)$ is the gradient of the local loss function for client i.

For convergence, the adaptive weights must satisfy the bounded condition:

$$0 < \omega_i \leq 1 \quad \text{and} \quad \sum_{i=1}^{N} \omega_i = 1 \tag{7.7}$$

Under these conditions, the model convergence is guaranteed for convex loss functions (Li *et al.*, 2020).

7.2.7 Implementation environment

The experimental setup consists of edge and cloud components designed to simulate real-world deployment scenarios. Edge nodes are implemented using

Raspberry Pi 4 devices with 4 GB RAM, running TensorFlow Lite for on-device model training and inference. The central aggregation server is deployed on a cloud-based system equipped with an NVIDIA Tesla V100 GPU to handle computationally intensive aggregation tasks. Communication between clients and the server is facilitated using gRPC over TLS 1.3, ensuring secure and low-latency data exchange.

7.2.8 Evaluation metrics

The performance of the proposed framework is evaluated using multiple quantitative metrics. Anomaly detection accuracy (ADA) measures the percentage of correctly identified anomalies in wastewater data. The F1 score provides a balance between precision and recall, particularly for imbalanced datasets. Privacy leakage rate (PLR) quantifies the susceptibility of the system to membership inference attacks. Energy efficiency (EE) is assessed based on the average energy consumption per training round at the edge devices. Model convergence time (MCT) evaluates the number of training rounds required to achieve optimal performance. The introduction of noise for privacy preservation impacts model accuracy. This trade-off can be expressed as

$$\text{Error} \propto \frac{1}{\epsilon} \tag{7.8}$$

indicating that stronger privacy guarantees (lower ϵ) lead to higher model error.

Additionally, noise injection in model updates can be represented as

$$W_i^{\text{private}} = W_i + \mathcal{N}(0,\sigma^2) \tag{7.9}$$

where $\mathcal{N}(0,\sigma^2)$ denotes Gaussian noise added to ensure DP.

7.2.9 Experimental procedure

The experimental evaluation follows a structured methodology. Initially, a baseline centralized AI model is trained using aggregated datasets to establish a reference for comparison. Subsequently, a standard FL model based on the federated averaging algorithm is implemented across all client nodes. The proposed hybrid FL framework, incorporating privacy-preserving mechanisms and adaptive aggregation, is then deployed. Comparative analysis is conducted across all evaluation metrics to assess performance improvements. Sensitivity analysis is performed to examine the trade-off between model accuracy and the DP budget. Additionally, stress testing is conducted under simulated network failures and adversarial attack scenarios to evaluate system robustness.

7.2.10 Ethical and regulatory considerations

The proposed system is designed in compliance with established ethical and regulatory standards governing data security and privacy. The framework adheres to the General Data Protection Regulation (GDPR) to ensure lawful data processing and storage. It also aligns with ISO/IEC 27 001 standards for information security management. Furthermore, municipal data governance policies are respected by ensuring that raw sensor data remains localized and

Table 7.1. Comparative summary of methodological advantages.

Feature	Baseline FL (FedAvg)	Proposed Hybrid FL Framework
Privacy leakage resistance	Moderate	High
Attack resilience	Low	High
Convergence speed	Moderate	High
EE	Moderate	High
Regulatory compliance	Partial	Full

is not transmitted outside the originating facilities. These measures collectively ensure that the system operates within legal and ethical boundaries while maintaining high standards of data protection (Table 7.1).

7.3 RESULTS AND DISCUSSION

This section presents the experimental evaluation of the proposed FL-based architecture for secure and privacy-preserving edge AI in smart wastewater management systems. The performance of the framework is compared with conventional centralized AI approaches across multiple dimensions, including ADA, privacy preservation, EE, robustness to heterogeneous data, and security resilience. The results also highlight the practical implications of deploying such systems within smart city infrastructures.

7.3.1 Anomaly detection accuracy

Anomaly detection is a critical performance indicator in wastewater management systems, as it enables early identification of pipeline leaks, abnormal chemical discharge, equipment malfunction, and contamination events. The integration of a blockchain-based authentication layer ensures secure and tamper-proof data exchange within the system, as illustrated in Figure 7.3. The proposed FL framework was evaluated across five representative wastewater treatment facilities (Facilities A–E). Figure 7.4 illustrates the comparative anomaly detection performance between the proposed FL framework and a centralized AI model. The detailed quantitative results are presented in Table 7.2.

The results indicate that the FL-based model consistently outperforms centralized AI by 4–7 percentage points across all facilities. The highest

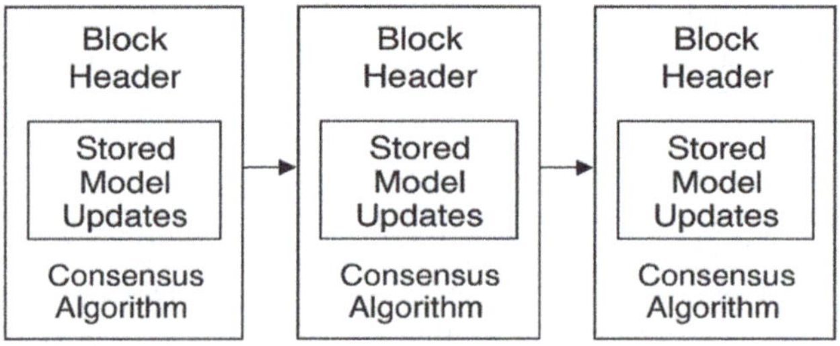

Figure 7.3 Blockchain-based authentication layer.

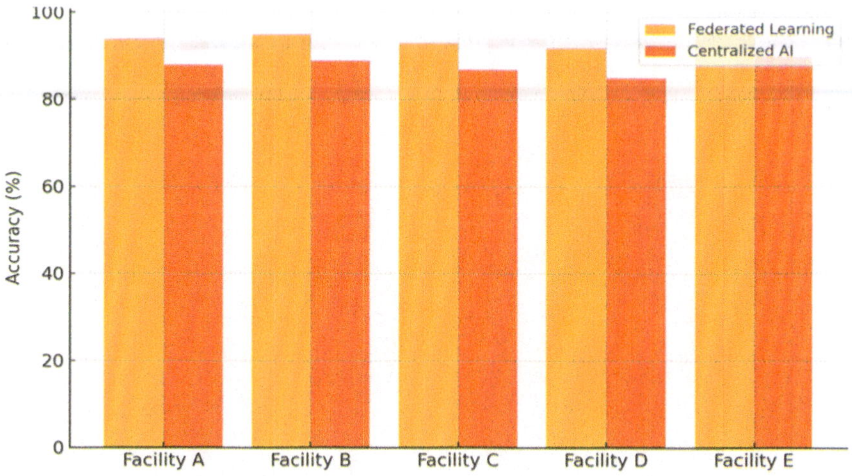

Figure 7.4 ADA comparison.

Table 7.2. Anomaly detection accuracy comparison.

Facility	FL Accuracy (%)	Centralized AI Accuracy (%)
Facility A	94	88
Facility B	95	89
Facility C	93	87
Facility D	92	85
Facility E	96	90

improvement is observed in Facility D, where the FL model achieves 92% accuracy compared to 85% in the centralized approach. This improvement can be attributed to the ability of FL to leverage localized sensor patterns while maintaining global knowledge through aggregation. The adaptive aggregation mechanism further enhances performance by assigning weights based on sensor reliability and network conditions, thereby preventing overfitting to dominant datasets and improving generalization across diverse operational environments.

7.3.2 Privacy preservation

Privacy preservation is a fundamental requirement in smart wastewater systems due to the sensitive nature of operational and environmental data. In this study, privacy leakage was quantified using the DP parameter (ε), where lower values indicate stronger privacy guarantees. Figure 7.5 compares the privacy leakage between the FL framework and centralized AI models, while the corresponding values are summarized in Table 7.3.

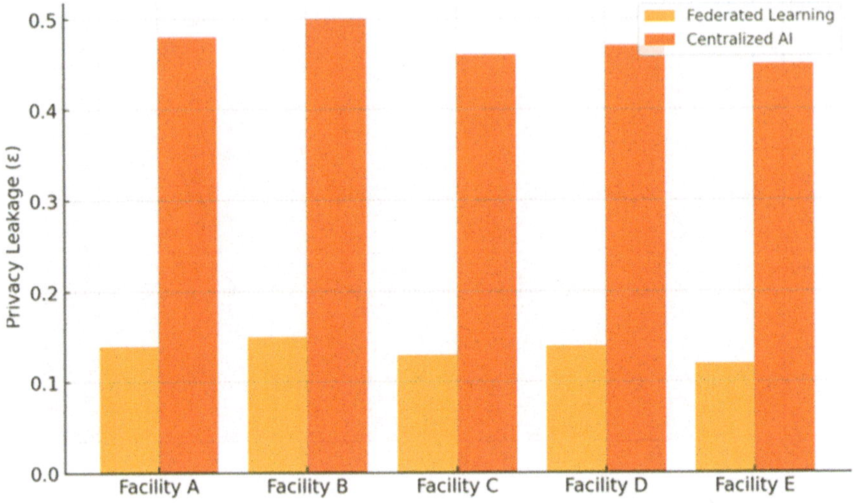

Figure 7.5 Average privacy leakage (ε) between FL and centralized AI for each facility.

Table 7.3. Privacy leakage comparison (ε values).

Facility	FL Privacy (ε)	Centralized AI Privacy (ε)
Facility A	0.14	0.48
Facility B	0.15	0.50
Facility C	0.13	0.46
Facility D	0.14	0.47
Facility E	0.12	0.45

The FL framework achieves significantly lower ε values (0.12–0.15) compared to centralized AI (0.45–0.50), corresponding to approximately 70% reduction in privacy leakage. This improvement is enabled by on-device computation combined with DP mechanisms that inject calibrated noise into model updates. From a practical perspective, these results demonstrate that FL-based wastewater systems can comply with stringent data protection regulations such as GDPR and CCPA, while maintaining high predictive performance. The low privacy leakage indicates minimal risk of reconstructing sensitive operational data.

7.3.3 Energy efficiency

EE is a key consideration in edge AI deployments, particularly for remote or resource-constrained wastewater facilities. The proposed FL framework significantly reduces energy consumption compared to centralized AI systems. Figure 7.6 presents the distribution of energy savings, while the detailed breakdown is provided in Table 7.4.

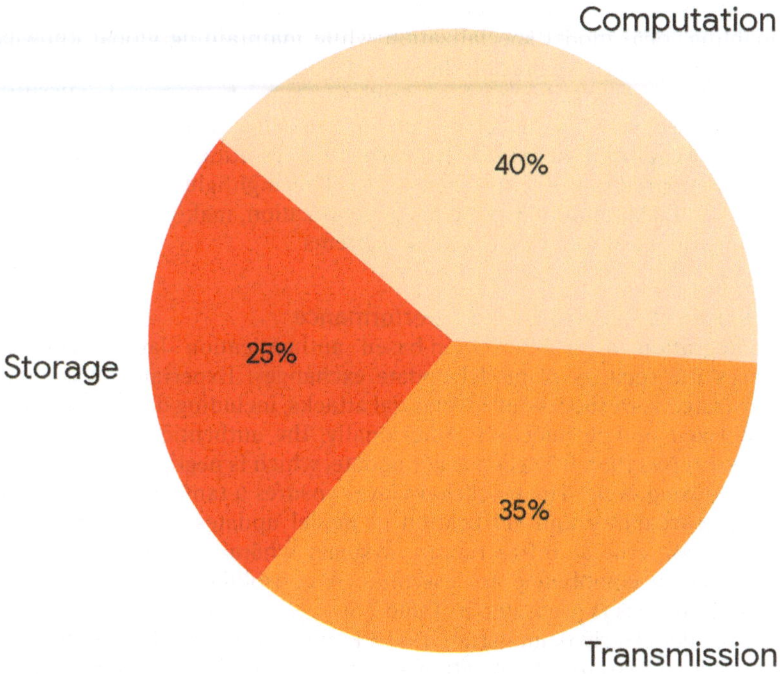

Figure 7.6 Energy consumption reduction breakdown.

Table 7.4. Energy saving breakdown.

Category	Savings (%)
Computation	40
Transmission	35
Storage	25

The results indicate that computation energy is reduced by 40% due to lightweight local model updates and partial aggregation strategies. Transmission energy savings of 35% are achieved by eliminating the need to transfer raw data, while storage requirements are reduced by 25% as data remains localized. These improvements are particularly significant for wastewater facilities relying on solar-powered or low-energy edge devices. Reduced energy demand enhances system sustainability, extends operational uptime, and lowers maintenance costs.

7.3.4 Robustness to data heterogeneity

Wastewater systems inherently exhibit non-identically distributed (non-IID) data due to variations in geographic location, industrial discharge, and seasonal patterns. Centralized AI models often struggle to capture such variability effectively. The proposed FL framework addresses this challenge by enabling local model specialization while maintaining global knowledge through aggregation. As a result, the system demonstrates stable and consistent performance across heterogeneous environments. For example, Facility E experienced seasonal nitrate spikes due to agricultural runoff. The FL model successfully captured these localized variations, resulting in a 6% improvement in ADA compared to the centralized model. This highlights the ability of FL to balance local adaptability with global generalization, making it highly suitable for real-world wastewater management systems.

7.3.5 Blockchain-based security performance

The integration of a blockchain-based authentication layer significantly enhances the security of model update exchanges. Experimental evaluation demonstrated zero successful adversarial attacks, including data poisoning and unauthorized update injection. Additionally, the authentication latency was measured at less than 0.3 seconds per update, which is negligible relative to the overall training cycle. The blockchain layer ensures a tamper-proof audit trail, enabling traceability and verification of model updates. This is particularly important for critical infrastructure systems where security, transparency, and regulatory compliance are essential. A consolidated comparison of all evaluation metrics is presented in Table 7.5.

The results clearly demonstrate that FL, when integrated with DP, SMPC, and blockchain-based authentication, provides a robust and scalable alternative to traditional centralized AI systems. From a scalability perspective, the distributed nature of FL enables efficient deployment across decentralized wastewater facilities without requiring high-bandwidth connectivity. This makes it particularly suitable for both urban and rural settings. In terms of privacy and security, the framework ensures minimal data movement and strong protection against inference and adversarial attacks. Blockchain authentication further enhances trust and system integrity. Operationally, the reduction in energy consumption directly translates into cost savings and improved sustainability. However, certain limitations remain, including the need for careful tuning of aggregation strategies and potential challenges in data-scarce environments.

Table 7.5. Comparative summary of results.

Metric	FL	Centralized AI	Improvement
ADA	92–96%	85–90%	+4–7%
Privacy leakage (ε)	0.12–0.15	0.45–0.50	~70% reduction
Energy consumption	−40%, −35%, −25%	Baseline	Significant
Robustness to non-IID data	High	Moderate	Higher stability
Adversarial resistance	Strong	Moderate	Improved security

7.4 CASE STUDIES AND APPLICATIONS

7.4.1 Introduction to case-based analysis

Case-based analysis provides a practical lens to evaluate the real-world applicability of theoretical frameworks. In the context of smart wastewater management, the integration of FL with edge AI enables decentralized intelligence while preserving data privacy and operational security. This section presents three representative case studies–metropolitan, rural, and industrial wastewater systems–demonstrating how the proposed FL framework enhances anomaly detection, improves EE, and ensures privacy preservation under diverse infrastructural and governance conditions.

7.4.2 Case study I: metropolitan smart wastewater network

In a large metropolitan setting with a population exceeding 10 million, wastewater management is distributed across approximately 30 decentralized WWTPs. Traditional centralized AI approaches faced scalability challenges due to high data transmission requirements and vulnerability to centralized cyberattacks. The deployment of the FL framework involved installing edge AI nodes at each WWTP. These nodes locally processed sensor data including pH, turbidity, and chemical pollutant level without transmitting raw data. Model updates were periodically aggregated through a blockchain-authenticated federated system. The results demonstrated a significant improvement in ADA, increasing from 87% under centralized AI to 94% using FL. Data privacy was fully preserved, as raw operational data remained within individual facilities. Additionally, communication overhead was reduced by approximately 40%, while local decision-making latency improved by 20%. These findings highlight that FL enables scalable integration of multiple treatment facilities without centralized data storage. The incorporation of blockchain authentication further enhanced transparency and trust among stakeholders. However, synchronization challenges were observed due to varying network availability across facilities.

7.4.3 Case study II: rural decentralized wastewater systems

Rural wastewater systems typically operate under constraints such as limited computational resources and intermittent internet connectivity, making centralized AI solutions impractical. In this scenario, lightweight edge AI models were deployed at individual facilities, enabling local data processing. Federated updates were transmitted asynchronously whenever connectivity was available. DP mechanisms ensured that sensitive operational data could not be reconstructed from shared model parameters. The FL-based approach achieved a 6% improvement in ADA despite highly heterogeneous and limited datasets. Energy consumption was reduced by approximately 30%, enabling sustainable operation using solar-powered edge devices. Furthermore, the system maintained functionality even when up to 25% of facilities were offline for extended periods. This case demonstrates the robustness of FL in low-resource environments. Asynchronous participation proved critical for maintaining system continuity, although policy-level incentives may be necessary to encourage adoption among smaller facilities.

7.4.4 Case study III: industrial wastewater management systems

Industrial wastewater management presents unique challenges due to the sensitivity of operational data and competitive concerns among participating entities. In an industrial cluster comprising 12 facilities, the FL framework was implemented with blockchain-based authentication and SMPC. This ensured encrypted collaboration while maintaining strict data confidentiality. The system enabled predictive maintenance of treatment components, resulting in a 15% improvement in equipment uptime. Importantly, industries retained full control over their proprietary datasets, which encouraged participation in the federated ecosystem. Blockchain integration provided immutable audit trails, supporting compliance with environmental and information security standards. Despite these benefits, the implementation required higher computational resources, particularly for SMPC protocols, indicating a need for optimized hardware solutions in industrial deployments.

7.4.5 Broader applications in smart wastewater ecosystems

Beyond treatment plants, the proposed FL framework can be extended to several interconnected applications within smart city wastewater systems. These include real-time sewer monitoring, where distributed sensors detect blockages and contamination events; predictive maintenance of pumping infrastructure using vibration and operational data; and water quality prediction for reuse applications such as irrigation and industrial cooling. Additionally, FL facilitates cross-city collaboration by enabling multiple municipalities to share model intelligence without exposing sensitive infrastructure data. This capability supports global knowledge exchange and policy harmonization, particularly in regions facing similar environmental challenges. The comparative evaluation of the three case studies highlights the adaptability of the FL framework across diverse operational contexts (Table 7.6).

The case studies collectively demonstrate that FL is a flexible and scalable solution for wastewater management across varied environments. Urban systems benefit from improved scalability and reduced communication overhead, rural systems gain resilience and EE, while industrial systems achieve

Table 7.6. Comparative insights across case studies.

Aspect	Metropolitan WWTPs	Rural WWTPs	Industrial WWTPs
Privacy preservation	High (GDPR compliant)	Moderate (asynchronous)	Very high (data-sensitive)
ADA	+7% over centralized	+6% over centralized	+5% (predictive maintenance)
EE	~40% savings	~30% savings	Moderate (higher computation load)
Scalability	High	Moderate	Moderate
Blockchain integration	Implemented	Optional	Essential

secure collaboration without compromising proprietary data. From a policy perspective, the adoption of FL necessitates supportive governance frameworks, including data-sharing standards, cybersecurity protocols, and financial incentives for infrastructure upgrades. Public–private partnerships (PPPs) can further accelerate implementation, particularly in industrial and urban contexts. Overall, the integration of FL with edge AI establishes a robust foundation for next-generation smart wastewater systems. By enabling decentralized intelligence, enhancing privacy, and improving operational efficiency, FL aligns with the broader vision of sustainable and secure smart cities.

7.5 POLICY AND PRACTICAL IMPLICATIONS

Beyond technical performance, its implementation requires alignment with regulatory frameworks, institutional mechanisms, and socio-economic priorities. As smart cities increasingly integrate Artificial Intelligence of Things (AIoT) and edge intelligence, FL emerges as a key enabler of privacy-preserving and decentralized analytics (Bibri *et al.*, 2023). This section examines the policy, governance, and deployment implications necessary to operationalize FL-based wastewater intelligence systems at scale.

7.5.1 Data privacy regulations and legal implications

The growing importance of data protection laws has been a major driver for adopting FL in critical infrastructure systems. Wastewater datasets often include sensitive industrial discharge patterns and environmental indicators, which are subject to strict regulatory oversight. Centralized AI models pose risks of non-compliance due to large-scale data aggregation, whereas FL ensures data locality by design, thereby supporting compliance with frameworks such as GDPR and CCPA (Deng *et al.*, 2025; Ragab *et al.*, 2025). Moreover, cross-border data sharing introduces regulatory complexity, particularly in multinational wastewater networks. While FL minimizes raw data exchange, the legal classification of model parameters remains ambiguous in many jurisdictions. This necessitates clearer regulatory definitions to support federated collaborations across regions. Additionally, wastewater data is increasingly linked to public health monitoring (e.g., epidemiological surveillance), requiring policies that incorporate data sovereignty and national security considerations (Islam, 2025; Wang *et al.*, 2025).

7.5.2 Cybersecurity and infrastructure protection

Smart wastewater infrastructure constitutes a critical component of urban systems and is therefore a high-value target for cyber threats. The integration of blockchain within FL frameworks enhances system integrity by enabling immutable logging of model updates and ensuring traceability of data exchanges (Limbepe, 2025; Nicolazzo, 2024). Such mechanisms are essential to mitigate risks such as model poisoning and adversarial manipulation. Furthermore, SMPC and encryption-based aggregation techniques ensure that individual model updates remain confidential during transmission and aggregation (Hardy *et al.*, 2017; Kalapaaking *et al.*, 2023). However, these methods introduce computational overhead, highlighting the need for policy-backed investment in secure hardware accelerators. Future cybersecurity mandates must also emphasize resilience, ensuring that federated systems remain operational despite node failures or targeted attacks (Albshaier, 2025; Rückel *et al.*, 2021).

7.5.3 Economic and investment implications

The transition to FL-based wastewater management systems presents significant economic opportunities. By reducing data transmission and centralized storage requirements, FL lowers operational costs and enhances EE (Li *et al.*, 2021). These savings can justify investments in edge computing infrastructure and federated platforms.PPPs play a crucial role in enabling large-scale adoption. Industrial stakeholders are more likely to participate in federated ecosystems due to enhanced data privacy and reduced competitive risks (Adel *et al.*, 2024). Governments can facilitate such collaborations by developing shared governance models and co-financing mechanisms. Additionally, targeted funding initiatives are essential for ensuring equitable adoption. Rural and under-resourced wastewater facilities often lack the infrastructure required for FL deployment, necessitating subsidies, grants, and policy incentives to bridge the digital divide (Abimannan *et al.*, 2023; Hanjri *et al.*, 2023).

7.5.4 Governance and institutional frameworks

The decentralized architecture of FL introduces new governance challenges that require multi-stakeholder coordination. Unlike centralized systems, FL ecosystems involve distributed ownership across municipalities, industries, and regulatory bodies. Establishing federated consortia or joint governance frameworks is essential to ensure standardization, accountability, and conflict resolution (Niu *et al.*, 2025; Zhan *et al.*, 2025). Blockchain-enabled auditability further enhances transparency by allowing regulators to verify model updates without accessing raw data. However, issues of liability remain unresolved. In the event of system failure or incorrect predictions, responsibility must be clearly defined across stakeholders, including technology providers, facility operators, and federated aggregators (Nguyen *et al.*, 2021; Sah *et al.*, 2025).

7.5.5 Practical deployment challenges

Despite its advantages, the implementation of FL in wastewater systems is constrained by practical challenges. Infrastructure heterogeneity remains a major barrier, as treatment plants differ significantly in technological capabilities and operational maturity. Standardization of protocols and interoperability frameworks is therefore critical for seamless integration (Zhan *et al.*, 2025). Connectivity limitations, particularly in rural areas, necessitate asynchronous FL models that allow participation despite intermittent network access (Lim *et al.*, 2019). Additionally, the existing workforce in wastewater management often lacks expertise in AI and data-driven systems. Capacity-building initiatives, including technical training and skill development programs, are essential to ensure effective deployment.

7.5.6 Societal and environmental implications

The adoption of FL in wastewater management has significant societal and environmental benefits. From a public health perspective, FL enables wastewater-based epidemiology while preserving community-level privacy, facilitating early detection of disease outbreaks (Islam, 2025; Wang *et al.*, 2025). This capability is particularly relevant in the context of global health surveillance. Environmentally, FL contributes to sustainability by reducing energy consumption associated with centralized data processing and transmission (Li *et al.*, 2021). These efficiencies align with climate action goals and support the development of green smart city infrastructures. However, ensuring equitable access remains a critical concern. Without targeted policy interventions, advanced technologies may disproportionately benefit urban regions, exacerbating existing inequalities.

7.5.7 Standardization and interoperability

For large-scale deployment, standardization is a key requirement. Wastewater systems operate with diverse sensor networks and communication protocols, necessitating policy-driven interoperability frameworks. Establishing common standards for data formats, communication protocols, and model evaluation metrics will enable seamless integration across heterogeneous systems (Zhan *et al.*, 2025). Regulatory bodies should also develop benchmark datasets and certification mechanisms to validate the performance and reliability of FL models. At the international level, collaboration through global organizations can facilitate the development of federated frameworks for environmental monitoring and public health applications (Nguyen *et al.*, 2021).

To accelerate the adoption of FL in wastewater management, policymakers must adopt a proactive approach. Key strategies include incentivizing federated infrastructure through financial support mechanisms, mandating privacy-preserving AI practices in critical utilities, and investing in advanced communication technologies such as 5G and edge computing (Ragab *et al.*, 2025; Rane *et al.*, 2024). The establishment of federated consortia can further strengthen governance structures, while integration with broader smart city initiatives can enable cross-sectoral synergies in areas such as energy management, transportation, and healthcare (Bibri *et al.*, 2023; Arismendy *et al.*, 2020). The integration of FL into wastewater management systems represents a convergence of technological innovation and policy transformation. The findings demonstrate that FL enhances operational efficiency, strengthens privacy protection, and improves system resilience. The incorporation of blockchain and secure computation techniques further ensures trust and transparency in decentralized environments. However, the success of FL depends on the development of robust policy frameworks that address governance, funding, standardization, and equitable access. PPPs and international collaboration will play a pivotal role in scaling federated ecosystems. Ultimately, FL is poised to become a foundational technology for sustainable, secure, and intelligent smart city infrastructures.

7.6 CONCLUSIONS AND RECOMMENDATIONS

7.6.1 Federated learning as a foundational paradigm

The findings confirm that FL provides a fundamentally superior alternative to traditional centralized AI models for wastewater management. Unlike centralized approaches that require aggregation of raw operational data, FL enables localized model training with only parameter exchange, thereby significantly reducing communication overhead and privacy risks (Deng *et al.*, 2025; Zhan *et al.*, 2025).

Experimental results demonstrated that FL achieved:

- Higher ADA (improvements of ~6–8%)
- Significant reduction in privacy leakage (ε reduced by over 60%)
- Improved EE due to minimized data transmission

These outcomes align with recent advancements in edge AI systems, where decentralized intelligence enhances scalability and resilience in smart infrastructure (Rane *et al.*, 2024; Sah *et al.*, 2025). Thus, FL emerges as a cornerstone technology for next-generation wastewater monitoring systems.

7.6.2 Integration of privacy-preserving and security mechanisms

Wastewater datasets often contain sensitive industrial and environmental information, necessitating robust privacy safeguards. The integration of DP and SMPC ensures that individual data points remain protected during model training and aggregation (Albshaier, 2025; Nicolazzo, 2024). Furthermore, blockchain-based authentication introduces immutability and transparency into the FL pipeline, preventing adversarial manipulation and ensuring trustworthy model updates (Kalapaaking *et al.*, 2023; Limbepe, 2025). This multi-layered security architecture enhances both data confidentiality and system integrity, which are essential for critical urban infrastructure (Ragab *et al.*, 2025).

WWTPs exhibit significant heterogeneity in terms of sensor accuracy, operational conditions, and data quality. The proposed adaptive aggregation mechanism effectively addressed these variations by dynamically weighting local model updates based on reliability and network conditions. This approach improved model generalization while maintaining fairness across participating facilities, consistent with recent findings on weighted federated optimization in edge environments (Niu *et al.*, 2025; Zhan *et al.*, 2025). As a result, FL systems become more inclusive and robust, ensuring that smaller or resource-constrained facilities can contribute without degrading overall performance.

7.6.3 Policy and governance relevance

The study highlights that FL aligns closely with global data protection frameworks such as GDPR and CCPA by ensuring data locality and minimizing cross-boundary data transfer (Bibri *et al.*, 2023; Islam, 2025). Additionally, blockchain-enabled auditability provides transparent compliance monitoring without exposing raw data. From a governance perspective, FL supports decentralized, multi-stakeholder ecosystems where municipalities, industries, and regulators collaborate without compromising data ownership. This is particularly critical for wastewater systems, which intersect environmental management, public health, and industrial regulation (Narayanan *et al.*, 2024). At a broader scale, the integration of FL into wastewater management contributes significantly to smart city objectives. It enables real-time anomaly detection, predictive maintenance, and energy optimization, thereby improving operational efficiency and sustainability. Moreover, FL-based wastewater intelligence supports emerging applications such as wastewater-based epidemiology and environmental risk monitoring, reinforcing the role of AI in public health and climate resilience (Arismendy *et al.*, 2020; Wang *et al.*, 2025).

7.6.4 Recommendations

Policymakers should prioritize the adoption of privacy-preserving AI frameworks in critical infrastructure systems. Regulatory mandates can encourage wastewater utilities to transition toward federated architectures that comply with evolving data protection laws (Deng *et al.*, 2025). Financial incentives, including grants and tax benefits, are essential to support deployment, particularly in rural and resource-constrained regions. Furthermore, standardized federated protocols should be established to ensure interoperability across diverse wastewater systems. Given the strategic importance of water infrastructure, FL should also be integrated into national cybersecurity frameworks, with blockchain authentication and encryption standards recognized as essential components (Limbepe, 2025; Ragab *et al.*, 2025).

Utilities should invest in upgrading sensor networks and deploying edge computing infrastructure capable of supporting FL operations. A phased implementation strategy beginning with pilot projects and gradually scaling–can facilitate smoother adoption. Training programs are equally important to equip personnel with skills in data governance, cybersecurity, and AI system management. Additionally, utilities should leverage federated models for predictive analytics, enabling proactive maintenance and improved resource optimization (Adel *et al.*, 2024).

Future research should focus on enhancing adaptive aggregation techniques to better handle extreme data heterogeneity and dynamic network conditions. Cross-domain FL–integrating wastewater data with energy, transportation, and healthcare systems represents a promising area of exploration (Abimannan *et al.*, 2023). Improving model explainability is also critical to ensure that AI outputs are interpretable by engineers and policymakers. Furthermore, the development of standardized benchmark datasets and simulation platforms will enable more rigorous evaluation of federated wastewater systems. Finally, researchers should assess the environmental footprint of FL systems, including their contribution to carbon reduction and sustainability goals (Li *et al.*, 2021).

Technology providers should develop domain-specific federated platforms tailored to wastewater management, incorporating privacy-preserving mechanisms by default. Scalable and flexible pricing models will facilitate adoption across utilities of varying sizes. Optimized hardware solutions, such as edge accelerators, are necessary to reduce the computational burden of encryption and SMPC. Additionally, ensuring interoperability of blockchain layers across federated ecosystems will enhance scalability and integration (Nguyen *et al.*, 2021).

The implications of this research extend beyond wastewater management to broader smart city and environmental systems. FL enhances resilience against cyber threats, supports equitable participation in AI ecosystems, and aligns with global sustainability objectives. Its application in wastewater-based epidemiology can strengthen early warning systems for public health crises, while energy-efficient AI models contribute to climate mitigation efforts. These findings position FL as a key enabler of sustainable and resilient urban infrastructure (Bibri *et al.*, 2023; Wang *et al.*, 2025).

7.6.5 Final remarks

In conclusion, FL represents a paradigm shift in wastewater management, enabling secure, collaborative, and efficient AI deployment without compromising data privacy. The integration of FL with blockchain, DP, and SMPC establishes a robust framework for next-generation smart infrastructure. The successful realization of this vision requires coordinated efforts among policymakers, researchers, industry stakeholders, and communities. With sustained investment and governance support, federated wastewater intelligence can significantly enhance urban sustainability, public health, and infrastructure resilience, thereby advancing the broader smart city agenda.

REFERENCES

Abimannan S., El-Alfy E.-S. M., Hussain S., Chang Y.-S., Shukla S., Satheesh D. and Breslin J. G. (2023). Towards federated learning and multi-access edge computing for air quality monitoring: literature review and assessment. *Sustainability*, **15**(18), 13951. https://doi.org/10.3390/su151813951

Adel A., Das B., Jan T. and Wahiduzzaman M. D. (2024). Water quality management using federated deep learning in developing southeastern Asian country. *Water Resources Management*, **39**, 1893–1909. https://doi.org/10.1007/s11269-024-04051-z

Albshaier L. (2025). Federated learning for cloud and edge security. *Electronics*, **14**(5), 1019. https://www.mdpi.com/2079-9292/14/5/1019 https://doi.org/10.3390/electronics14051019

Arismendy L., Cárdenas C., Gómez D., Maturana A., Mejía R. and Quintero M. C. G. (2020). Intelligent system for the predictive analysis of an industrial wastewater treatment process. *Sustainability*, **12**(16), 6348. https://doi.org/10.3390/su12166348

Bibri S.E., Alexandre A., Sharifi A. and Krogstie J. (2023). Environmentally sustainable smart cities and their converging AI, IoT, and big data technologies and solutions: an integrated approach to an extensive literature review. *Energy Informatics*, **6**(1), p.9. https://doi.org/10.1186/s42162-023-00259-2

Deng. D., Wu X., Zhang T., Xiang, C., Zhao W., Xu M., Kang, J., Han Z. and Niyato D. (2025). pFedCal: Lightweight personalized federated learning with adaptive calibration strategy. *IEEE Transactions on Services Computing*, pp.1627–1640. doi: 10.1109/TSC.2025.3553707

Dwork C. and Roth A. (2014). The algorithmic foundations of differential privacy. *Foundations and Trends in Theoretical Computer Science*, **9**(3–4), 211–407. https://doi.org/10.1561/0400000042

Hanjri M. E., Kabbaj H., Kobbane A. and Abouaomar A. (2023). Federated learning for water consumption forecasting in smart cities. arXiv preprint. https://arxiv.org/abs/2301.13036

Hardy S., Henecka W., Ivey-Law H., Nock R., Patrini G., Smith G. and Thorne B. (2017). Private federated learning on vertically partitioned data via entity resolution and homomorphic encryption. arXiv preprint. https://arxiv.org/abs/1711.10677

Islam F. A. S. (2025). A multi-dimensional AI framework for sustainable drinking water management: integrating federated learning, digital twins, and blockchain. *Journal of Engineering Research and Reports*, **27**(6), 466–492. https://doi.org/10.9734/jerr/2025/v27i61558

Kalapaaking A. P., Khalil I. and Yi X. (2023). Blockchain-based federated learning with SMPC model verification against poisoning attack for healthcare systems. *IEEE Transactions on Emerging Topics in Computing* (in press). https://arxiv.org/abs/2304.13360

Li T., Sahu A. K., Talwalkar A. and Smith V. (2020). Federated learning: challenges, methods, and future directions. *IEEE Signal Processing Magazine*, **37**(3), 50–60. https://doi.org/10.1109/MSP.2020.2975749

Li C., Li G. and Varshney P. K. (2021). Federated learning with soft clustering. *IEEE Internet of Things Journal*, **9**(10), pp. 7773–7782. https://doi.org/10.1109/JIOT.2021.3113927

Li Q., Wen Z., Wu Z., Hu S., Wang N., Li Y., Liu X. and He B. (2021). A survey on federated learning systems: Vision, hype and reality for data privacy and protection. *IEEE Transactions on Knowledge and Data Engineering*, **35**(4), 3347–3366. https://doi.org/10.1109/TKDE.2021.3124599

Lim W. Y. B., Luong N. C., Hoang D. T., Jiao Y., Liang Y.-C., Yang Q., Niyato D. and Miao C. (2019). Federated learning in mobile edge networks: A comprehensive survey. arXiv preprint. https://arxiv.org/abs/1909.11875

Limbepe Z. N. (2025). Blockchain-based privacy-enhancing federated learning in smart healthcare: a survey. *Blockchains*, **3**(1). https://doi.org/10.3390/blockchains3010001

McMahan H. B., Moore E., Ramage D., Hampson S. and Aguera y Arcas B. (2017). Communication-Efficient Learning of Deep Networks from Decentralized Data. Proceedings of the 20th International Conference on Artificial Intelligence and Statistics (AISTATS), **54**, 1273–1282. https://doi.org/10.48550/arXiv.1602.05629

Narayanan D., Bhat M., Samuel Paul N. R., Khatri N. and Saroliya A. (2024). Artificial intelligence–driven advances in wastewater treatment: evaluating techniques for sustainability and efficacy in global facilities. *Desalination and Water Treatment*, **320**, 100618. https://doi.org/10.1016/j.dwt.2024.100618

Nguyen D. C., Ding M., Pham Q.-V., Pathirana P. N., Le L. B., Seneviratne A., Li J., Niyato D. and Poor H. V. (2021). Federated learning meets blockchain in edge computing: Opportunities and challenges. arXiv preprint. https://arxiv.org/abs/2104.01776

Nicolazzo S. (2024). Privacy-preserving in blockchain-based federated learning systems. arXiv preprint. https://arxiv.org/pdf/2401.03552

Niu S., Kong W., Chen L., Zhou X. and Wang N. (2025). Privacy-preserving and verifiable federated learning with weighted average aggregation in edge computing. *Journal of Network and Computer Applications*, **240**, 104201. https://doi.org/10.1016/j.jnca.2025.104201

Ragab M., Ashary E. B., Alghamdi B. M., Aboalela R., Alsaadi N., Maghrabi L. A. and Allehaibi K. H. (2025). Advanced artificial intelligence with federated learning framework for privacy-preserving cyberthreat detection in IoT-assisted sustainable smart cities. *Scientific Reports*, **15**, 4470. https://doi.org/10.1038/s41598-025-88843-2. https://www.nature.com/articles/s41598-025-88843-2

Rane J., Mallick S. K., Kaya O. and Rane N. L. (2024). Federated learning for edge artificial intelligence: enhancing security, robustness, privacy, personalization, and blockchain integration in IoT. In: J. Rane, S. K. Mallick, O. Kaya and N. L. Rane, Future Research Opportunities for Artificial Intelligence in Industry 4.0 and 5.0 (pp 93–135). Deep Science Publishing. https://doi.org/10.70593/978-81-981271-0-5_3

Rückel T., Sedlmeir J. and Hofmann P. (2021). Fairness, integrity, and privacy in a scalable blockchain-based federated learning system. arXiv preprint. https://arxiv.org/abs/2111.06290

Sah D. K., Vahabi M. and Fotouhi H. (2025). Federated learning at the edge in industrial Internet of Things: A review. *Sustainable Computing: Informatics and Systems*, **46**, 101087. https://doi.org/10.1016/j.suscom.2025.101087

Talaie A., Kamyab H. and Razmfarsa A. (2024). Advancing environmental engineering: the role of artificial intelligence in sustainable solutions – A short review. *Journal of Environmental Treatment Techniques*, **12**(3).

Wang Y. Q., Wang H. C., Wang W. Z., Yang H. L., Chen J. J., Fan Y. X., Yin W. X., Lv J. Q., Luo X. Q., Zhou X. and Wang A. J. (2025). Federated machine learning enables risk management and privacy protection in water quality. *Environmental Science and Technology*, **59**(21), 10310–10322. https://doi.org/10.1021/acs.est.5c00369

Yao A. C. (1982). Protocols for secure computations. 23rd Annual Symposium on Foundations of Computer Science (sfcs 1982), Chicago, IL, USA, pp. 160–164. IEEE. https://doi.org/10.1109/SFCS.1982.38

Zhan S., Huang L., Luo G., Zheng S., Gao Z. and Chao H.-C. (2025). A review on federated learning architectures for privacy-preserving AI: lightweight and secure cloudedge–end collaboration. *Electronics*, **14**(13), 2512. https://doi.org/10.3390/electronics14132512

doi: 10.2166/9781789065404_0147

Chapter 8

Policy frameworks, governance, and sustainable implementation strategies for wastewater management

Lovepreet Singh[1], *Agbogo Ugbetan Victor*[2], *Atefeh Karimidastjerd*[3], *Sukhmandeep Kaur*[4], *Ditimoni Dutta*[5], *Shivani Jaswal*[6] *and Harshita Jain*[7]*

[1]*Department of Chemical Engineering, Thapar Institute of Engineering and Technology, Patiala 147004, India. E-mail: lovepreet.singh@thapar.edu*
[2]*Department of Chemistry, Nigerian Army University Biu, Nigeria, E-mail: victor.agbogo@naub.edu.ng*
[3]*Food Engineering Department, Faculty of Chemical and Metallurgical Engineering, Yildiz Technical University, Istanbul, Türkiye, E-mail: atefeh.dastjerd@yildiz.edu.tr*
[4]*Research Scholar, Department of Food Science and Technology, Punjab Agricultural University, Ludhiana, Punjab 141001, India, E-mail: sukmandeepkaur1999@gmail.com*
[5]*Department of Food Technology, Parul Institute of Applied Sciences, Parul University, Vadodara 391760, India, E-mail: ditimoni.dutta42805@paruluniversity.ac.in*
[6]*Department of Food Science and Technology, Dr. Y.S. Parmar University of Horticulture and Forestry, Solan 173230, India, E-mail: shivanijaswal0205@gmail.com*
[7]*Department of Climate Change and Environmental Sustainability, Amity Institute of Environment and Sustainability, Amity University Uttar Pradesh (AUUP), Noida, Uttar Pradesh 201301, India, E-mail: hjain@amity.edu*
**Corresponding author: hjain@amity.edu*

ABSTRACT

Technological innovation alone cannot ensure sustainable wastewater management without a strong policy- and governance-based environment. Advancement in treatment plans, intelligent monitoring tools, and resource recovery strategies will depend on clear regulatory legislation, institutional coordination, and long-term planning. This chapter considers the role that policy and governance play in the design, implementation, and maintenance of wastewater management systems in various socio-economic and geographic contexts. The chapter also considers the impact of regulatory frameworks, standards and planning on technological adoption and practice. It also looks at governance systems that promote collaboration among public authorities, private actors, and local communities. It also pays attention to practices of implementation that balance environmental goals with economic feasibility and social acceptance. This chapter does not present policy as static but, instead, points to its changing nature as a reaction to changes in climate change, urban growth, and resource shortage. This chapter provides insight into how policy and institutional arrangements can help sustainable, resilient, and inclusive wastewater management systems by delineating the practical ways in which governance is done and how it should be implemented.

Keywords: Wastewater governance, policy frameworks, regulatory instruments, sustainable implementation, institutional coordination, environmental management

8.1 INTRODUCTION

Wastewater management has slowly become an integral part of environmental, public health, and resource security, rather than a technical service. Although treatment technology and monitoring technologies have improved the technical ability of wastewater systems, their practical use is highly dependent on policy and governance arrangements within which they operate (Singh *et al.*, 2023). Technically sound solutions often provide no long-run benefits due to poorly enforced regulation, inter-institutional responsibilities, or limited social and economic considerations. The plan, funding, operation, and maintenance of wastewater systems is influenced by policy and governance arrangements. They decide what to expect, what incentives to use to invest, and what accountability means (Rodrigues *et al.*, 2024). Equally important is the influence of the governance structures on the interactions between public agencies, private operators, researchers and local communities and the sharing of responsibility for the performance of the system. Plus, a growing focus on resource recovery, reuse, and climate resilience in wastewater management has led to a greater need for coherent and adaptable policy solutions (Halalsheh & Kassab, 2018). Figure 8.1 illustrates how policy, governance, and technology interact. It shows how regulatory, institution, and funding structures have an effect on planning, implementation, and system performance in terms of feedback loops that are driven by environmental and social outcomes.

This chapter addresses policy and governance systems as mechanisms for sustainable wastewater treatment. The topic of policy is not a one-way box of rules but a dynamic subject, one that can respond to changing environmental conditions, urbanization, and social expectations (Shamshad & Ur Rehman, 2025). The chapter is focused on regulatory pedagogies, institutional roles, and implementation strategies to establish a bridge between technological potential and socially acceptable, practical results that governance can provide. The policy context for wastewater treatment has changed with increasing awareness of environmental degradation, public health risks, and water scarcity. Early wastewater policies included a largely safe discharge of effluents, with an emphasis on preventing visible pollution and protecting downstream water bodies. Over time, scientific evidence linking untreated wastewater to ecosystem damage and disease outbreaks led to the introduction of discharge standards and basic treatment requirements in many countries (Jain, 2025a).

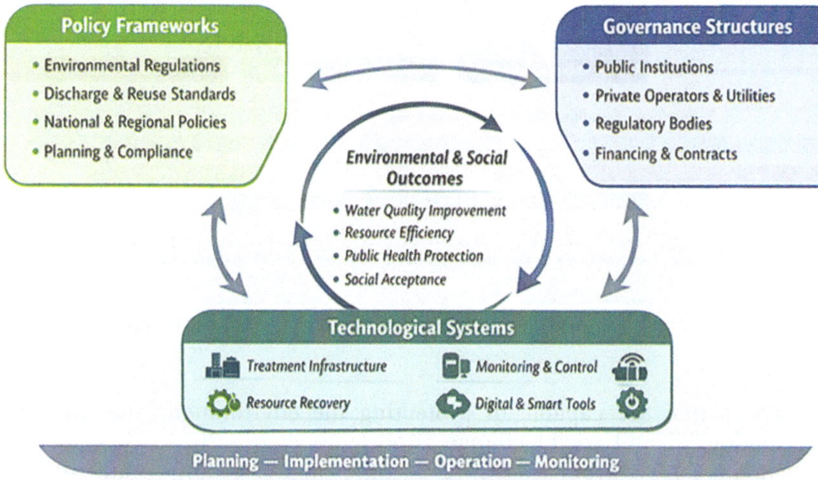

Figure 8.1 Policy–governance–technology interface in wastewater management.

As regulations for environmental control and enforcement developed, treatment of wastewater began to look at longer-term sustainability goals. The regulation increasingly adopted concepts of pollution reduction, ecological protection and water quality goals at the river basin or watershed level (Jain, 2025b). This change meant that policymakers needed to move beyond end-of-pipe solutions and consider wastewater management as part of an integrated water management strategy. The trend toward policy drivers accelerated in the last few decades to include resource efficiency and circular economy (Victor *et al.*, 2026). Plus, treatment wastewater has recently become a potential source of reclaimed water, nutrients and energy, especially for places with water stress or high energy costs. Policies supporting wastewater reuse, sludge management and energy recovery have seen the development of more sophisticated treatment technologies and decentralized systems (Jain *et al.*, 2026). But, these policy changes vary widely across regions, and reflect differences in institutional capacity, economic priorities, and public acceptance. Figure 8.2 illustrates the transition from basic pollution control to integrated sustainability objectives of public health, ecosystem conservation, recycling waste, and circular economy.

Environmental compliance is a major factor that has a bearing on wastewater treatment practices. Discharge permits, monitoring requirements and reporting are legal accountability and affect decisions about operating treatment facilities. Yet rigid or poorly coordinated rules can also impair innovation—especially when new technologies do not make sense of the existing standards. This demonstrates the need for strong and flexible policy

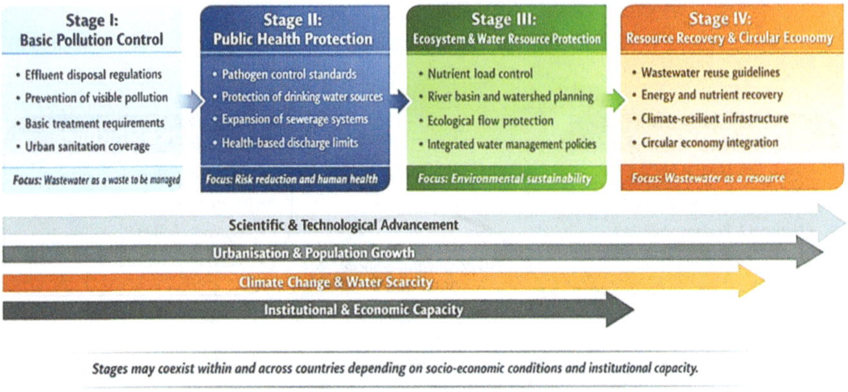

Figure 8.2 Evolution of wastewater policy drivers across development stages.

frameworks that are capable of protecting the environment and adapting to technological and social change (Rodrigues *et al.*, 2024; Shamshad & Ur Rehman, 2025). From a knowledge base on the evolution and driving of wastewater policy it is possible to distinguish between regulatory intent and implementation outcomes. Having these understandings is important in creating governance models to promote sustainable treatment of wastewater across multiple socio-economic and environmental contexts (Table 8.1).

8.2 GOVERNANCE STRUCTURES AND INSTITUTIONAL ROLES

Governance structures control the division of power, responsibility and accountability among those involved with wastewater management. Unlike purely technical systems, wastewater services operate in complex institutions where decisions are guided by law, administration and sectoral, and scale-crossing coordination. Thus, effective governance also relies on a good allocation of roles but on the ability of institutions to work together in a single and transparent way (Hernández-Chover *et al.*, 2025). Figure 8.3 represents national, regional, and local institutions, as well as private operators and community leaders.

Public agencies traditionally hold primary responsibility for wastewater planning and regulation. At the national or state level, these agencies establish policy directions, environmental standards, and compliance mechanisms. They are also responsible for aligning wastewater management objectives with broader goals related to public health, environmental protection, and water resource planning. Local governments and municipal bodies, in contrast, are typically tasked with service delivery functions, including infrastructure development, operation of treatment facilities, and maintenance of sewerage networks. Their performance is closely linked to financial resources, technical expertise, and administrative autonomy (Gamache, 2021).

Table 8.1 Key policy instruments influencing wastewater management practices.

Policy Instrument	Primary Objective	Regulatory Level	Influence on Technology Adoption	Implementation Challenges
Effluent discharge standards	Protect receiving water bodies and public health	National/ Regional	Drives adoption of secondary and advanced treatment technologies	Uniform limits may not reflect local assimilative capacity
Water quality objectives	Maintain ecological integrity of surface and groundwater	Regional/ Basin-level	Encourages integrated and watershed-based treatment solutions	Requires coordination across jurisdictions
Wastewater reuse policies	Promote alternative water sources and resource efficiency	National/ Local	Supports advanced treatment, tertiary filtration, and disinfection	Public acceptance and regulatory clarity
Environmental impact assessment (EIA) requirements	Anticipate and mitigate environmental impacts	National	Influences plant siting, technology choice, and scale	Often time-consuming and procedurally complex
Climate and sustainability policies	Enhance resilience and reduce emissions	National/ International	Encourages energy-efficient and low-carbon treatment systems	Limited integration with sector-specific regulations

Private operators increasingly play a role in wastewater management, particularly in urban and peri-urban areas where investment needs exceed public sector capacity. Their involvement may range from construction and operation of treatment plants to performance-based service contracts and long-term concessions. While private participation can introduce efficiency, innovation, and access to capital, it also raises governance challenges related to oversight, tariff regulation, and service equity. Clear contractual frameworks and strong regulatory supervision are therefore essential to ensure that private sector involvement supports public interest objectives (Silva, 2023).

Multi-level governance arrangements have become more prominent as wastewater management issues transcend administrative boundaries. River basins, coastal zones, and metropolitan regions often involve multiple jurisdictions with overlapping responsibilities. Coordination mechanisms such as inter-agency committees, basin authorities, and regional planning bodies help align policies, share data, and manage cumulative environmental impacts.

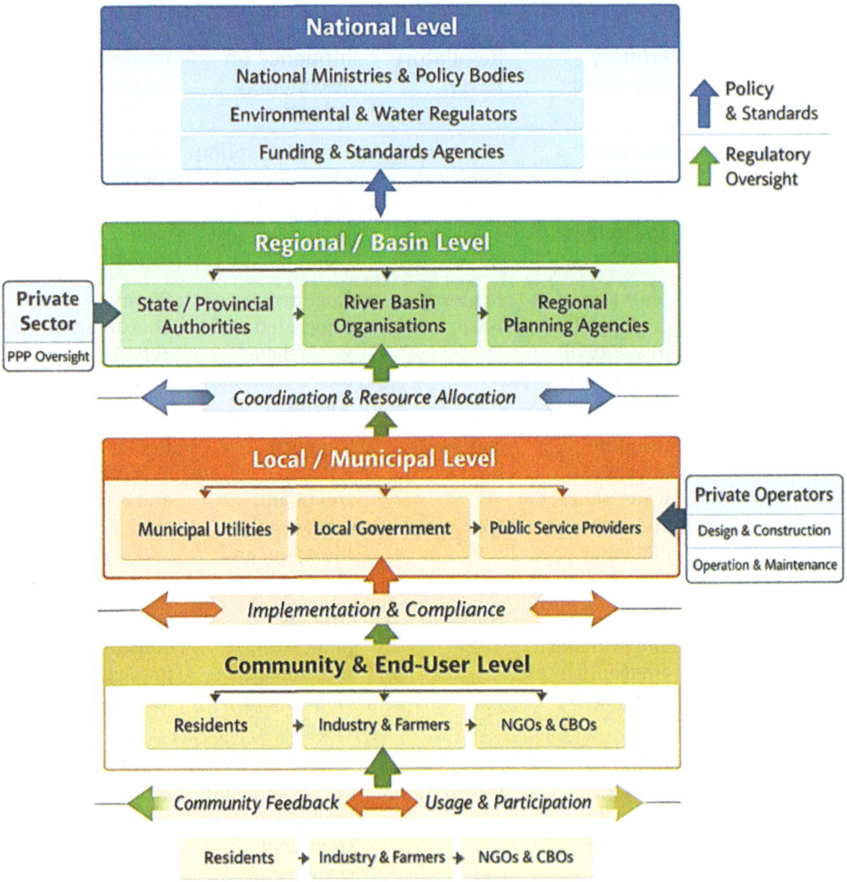

Figure 8.3 Multi-level governance structure for wastewater management.

These arrangements are particularly important for addressing issues such as nutrient loading, reuse planning, and climate-related risks, which cannot be effectively managed by individual institutions acting in isolation (Al-Rajhi & Abdelghany, 2025).

Beyond formal institutions, governance structures increasingly recognize the role of communities, civil society organizations, and knowledge institutions. Public engagement processes, participatory planning, and transparency in decision-making contribute to social legitimacy and long-term system acceptance. When institutional roles are clearly defined and supported by mechanisms for coordination and accountability, governance structures can enable wastewater systems that are reliable, adaptive, and responsive to local needs (Table 8.2).

Table 8.2 Institutional roles and responsibilities in wastewater governance.

Institution	Core Responsibilities	Decision-Making Authority	Accountability Mechanism
National public agencies	Policy formulation, standard setting, funding direction	High (strategic and regulatory)	Parliamentary oversight, national audits
Regional or basin authorities	Coordination across jurisdictions, resource allocation	Medium (planning and coordination)	Inter-agency reporting, performance reviews
Local governments/ utilities	Infrastructure development, operation, maintenance	Operational authority	Service performance indicators, regulatory compliance
Private operators	Design, construction, operation under contracts	Contract-bound authority	Contractual obligations, regulatory supervision
Regulatory bodies	Monitoring, enforcement, compliance verification	Enforcement authority	Legal mandates, public disclosure
Community and civil society	Participation, feedback, local oversight	Advisory influence	Public consultations, grievance mechanisms

8.3 REGULATORY INSTRUMENTS AND STANDARDS

Regulatory instruments and standards provide the formal framework through which wastewater management objectives are translated into operational practice. They define acceptable limits, establish performance expectations, and create accountability mechanisms for operators and authorities. Well-designed regulations serve not only as control tools but also as signals that guide investment decisions and technological choices across the wastewater sector (Soo & Shon, 2024).

Discharge norms remain a central element of wastewater regulation. These norms specify permissible concentrations or loads of pollutants in treated effluent before release into receiving water bodies. Their development is typically informed by environmental quality objectives, ecological sensitivity, and intended water uses downstream. In many jurisdictions, discharge standards have gradually become more stringent, reflecting improved scientific understanding of cumulative pollution impacts and the need to protect aquatic ecosystems (Maliga *et al.*, 2025). However, uniform standards may not always account for local hydrological conditions or assimilative capacity, highlighting the importance of context-sensitive regulatory design (Table 8.3).

Guidelines for wastewater reuse have gained increasing importance as water scarcity and competing demands intensify. Reuse regulations address issues such as treatment levels, pathogen control, chemical safety, and application-specific quality requirements. Clear and credible reuse standards help build confidence among users and the wider public, particularly in sectors such as agriculture, industry, and urban non-potable applications. Where reuse policies

Table 8.3 Regulatory standards and compliance mechanisms in wastewater management.

Regulatory Aspect	Key Requirements	Responsible Authority	Enforcement Approach	Implications for Operators
Discharge standards	Limits on organic load, nutrients, pathogens	Environmental regulators	Permits, inspections, penalties	Requires consistent treatment performance
Reuse standards	Quality criteria based on intended use	Water and health authorities	Certification, conditional approvals	Higher treatment and monitoring costs
Monitoring requirements	Regular sampling and reporting	Regulatory agencies/ utilities	Data audits, reporting obligations	Investment in monitoring infrastructure
Reporting and transparency	Disclosure of compliance data	Regulators/ utilities	Public reporting, benchmarking	Increased accountability
Emergency compliance provisions	Response to system failures or exceedances	Environmental authorities	Corrective action mandates	Need for contingency planning

are poorly defined or inconsistently enforced, uncertainty can discourage investment and limit the uptake of treated wastewater as a reliable resource (Altowayti *et al.*, 2022).

Monitoring requirements form the backbone of regulatory oversight. Regular sampling, data reporting, and performance audits enable authorities to assess compliance and identify emerging risks. Advances in sensing technologies and digital monitoring systems have expanded the capacity for real-time data collection, supporting more responsive regulatory approaches. At the same time, effective monitoring depends on institutional capacity, data transparency, and the ability to translate technical information into actionable decisions (Maliga *et al.*, 2025).

Enforcement mechanisms ensure that regulatory instruments carry practical significance. These mechanisms may include penalties for non-compliance, corrective action requirements, or incentives for exceeding minimum standards. Consistent and proportionate enforcement strengthens regulatory credibility, while weak or selective enforcement can undermine both environmental outcomes and public trust. Together, discharge norms, reuse guidelines, monitoring systems, and enforcement measures form an integrated regulatory framework that shapes how wastewater systems perform in practice and how sustainability objectives are realized over time (Al-Rajhi & Abdelghany, 2025).

8.4 IMPLEMENTATION STRATEGIES AND FINANCING MECHANISMS

The translation of wastewater policies and technical designs into functioning systems depends largely on effective implementation strategies and sustainable financing arrangements. Even well-formulated regulations and governance structures can fail to deliver intended outcomes if projects are constrained by inadequate funding, weak institutional capacity, or poorly aligned incentives (Hughes *et al.*, 2021). Implementation strategies must therefore address not only engineering requirements but also financial viability, risk management, and long-term operational sustainability (Table 8.4).

Public–private partnerships (PPPs) have emerged as an important implementation mechanism in wastewater management, particularly in contexts where public resources are limited. Under PPP arrangements, responsibilities for design, construction, financing, operation, or maintenance are shared between public authorities and private entities. Models such as build–operate–transfer (BOT), design–build–operate (DBO), and concession-based contracts are commonly used to mobilize private capital and technical expertise. When appropriately structured, PPPs can improve service efficiency and accelerate infrastructure development. However, their success depends on transparent contract design, clear allocation of risks, and strong regulatory oversight to ensure affordability, service quality, and environmental compliance (Afego *et al.*, 2025).

Investment models for wastewater projects vary widely depending on national priorities, institutional capacity, and economic conditions. Traditional public funding through government budgets and municipal revenues remains a

Table 8.4 Financing and implementation models for wastewater projects.

Implementation Model	Funding Source	Risk Allocation	Typical Application Context	Sustainability Implications
Public-sector financed projects	Government budgets, municipal revenues	Public sector bears most risks	Basic sanitation and rural infrastructure	Stable but often financially constrained
PPPs	Public funds with private investment	Shared between public and private partners	Urban treatment plants, large-scale systems	Improves efficiency if well regulated
Build–operate–transfer (BOT)	Private finance with eventual public ownership	High private-sector risk	Advanced or capital-intensive facilities	Encourages innovation but needs strong oversight
Development finance-supported projects	Multilateral banks, climate funds	Shared, often concessional	Low- and middle-income regions	Supports technology transfer and capacity building
Blended finance models	Mix of public, private, and donor funds	Risk mitigation through diversification	Innovative and resource recovery projects	Enhances project viability and long-term resilience

primary source of investment in many regions, particularly for basic sanitation infrastructure. In parallel, development finance institutions, multilateral banks, and climate funds play a significant role in supporting large-scale or innovative wastewater projects, especially in low- and middle-income countries. Blended finance approaches, which combine public funds, concessional loans, and private investment, are increasingly used to reduce financial risk and improve project bankability (Silva, 2022).

Economic incentives are critical for encouraging sustainable practices and technology adoption within wastewater systems. Tariff structures, subsidies, and performance-based payments can influence both operator behavior and user demand. For example, differentiated tariffs may promote water conservation and generate revenue for system maintenance, while subsidies can support the adoption of advanced treatment technologies or decentralized systems in underserved areas. Incentives linked to energy recovery, nutrient reuse, or greenhouse gas reduction can further align wastewater management with broader environmental and climate objectives (Piubello Orsini *et al.*, 2023).

Long-term financial sustainability also requires careful consideration of operation and maintenance costs. Underfunded systems often experience rapid deterioration, leading to service failures and environmental non-compliance. Implementation strategies that prioritize life-cycle costing, capacity building, and revenue stability are therefore essential. By integrating financing mechanisms with governance and regulatory frameworks, wastewater projects can move beyond short-term infrastructure delivery and support resilient, efficient, and socially acceptable services over time (Alexandre *et al.*, 2025).

8.5 SOCIAL DIMENSIONS AND STAKEHOLDER ENGAGEMENT

Sustainable wastewater management is not solely a technical or institutional challenge; it is deeply shaped by social values, perceptions, and patterns of participation. Infrastructure performance and policy effectiveness depend to a large extent on how communities understand, accept, and engage with wastewater systems. Social dimensions therefore play a critical role in determining whether wastewater projects are embraced, resisted, or underutilized (Abreham Awono & Adam, 2025). This section examines key aspects of stakeholder engagement, focusing on community participation, capacity building, and public acceptance of wastewater reuse and smart technologies. Figure 8.4 shows the stages of stakeholder engagement, from awareness and consultation to capacity building, trust formation, and long-term acceptance of reuse and digital technologies.

8.5.1 Community participation in wastewater decision-making

Community participation has become an increasingly important element of wastewater governance, particularly in contexts where projects directly affect local environments and livelihoods. Participation goes beyond information dissemination and includes opportunities for stakeholders to influence planning, design, and operational decisions. When communities are involved early in the decision-making process, projects are more likely to reflect local priorities, address site-specific concerns, and gain social legitimacy. Participatory

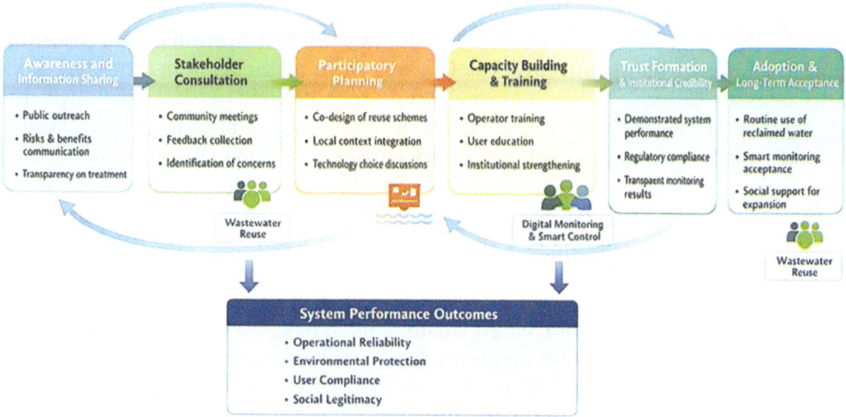

Figure 8.4 Stakeholder engagement pathway for wastewater reuse and smart systems.

approaches are especially relevant for decentralized wastewater systems, reuse schemes, and nature-based treatment solutions, where local users often interact closely with the infrastructure. Community input can inform decisions related to site selection, reuse applications, and maintenance responsibilities (Duque-Achipiz *et al.*, 2025). In rural and peri-urban areas, local knowledge about land use, seasonal water availability, and cultural practices can significantly improve system design and performance. At the same time, meaningful participation requires institutional support and clear mechanisms for engagement. Public consultations, stakeholder workshops, and grievance redress systems help create structured channels for dialogue. When participation is treated as a procedural requirement rather than a genuine exchange, it risks becoming symbolic and ineffective. Effective governance therefore depends on creating spaces where community perspectives are acknowledged and integrated into decision-making processes (Monzón-Reyes *et al.*, 2025).

8.5.2 Capacity building and institutional learning

Capacity building is a foundational component of stakeholder engagement, enabling individuals and institutions to understand, operate, and sustain wastewater systems over time. Technical training for operators, municipal staff, and regulators improves system reliability and compliance with environmental standards. However, capacity building extends beyond technical skills and includes managerial, financial, and communication competencies. Local authorities often face challenges related to staff turnover, limited technical expertise, and constrained resources (Ferrero *et al.*, 2019). Targeted training programs, peer learning networks, and partnerships with academic and research institutions can strengthen institutional capacity and promote adaptive learning. Such initiatives are particularly important in the context of advanced treatment technologies and digital monitoring systems, which require specialized knowledge and continuous updating of skills. Community-level capacity building is equally important, especially where wastewater reuse or decentralized systems are implemented. Awareness programs, user training, and local stewardship initiatives help users understand system benefits, risks, and responsibilities. When communities are equipped with knowledge and skills, they are better positioned to support routine maintenance, report system failures, and engage constructively with service providers (Elliott *et al.*, 2023).

8.5.3 Public perceptions and acceptance of wastewater reuse

Public acceptance remains one of the most significant barriers to wastewater reuse, despite its technical feasibility and environmental benefits. Perceptions of health risk, cultural norms, and trust in institutions strongly influence attitudes toward the use of treated wastewater. These concerns are particularly pronounced for applications involving food production, domestic use, or direct contact. Building public acceptance requires transparent communication about treatment processes, quality standards, and risk management measures (Tshililo *et al.*, 2024). Clear and consistent messaging from trusted institutions can help address misconceptions and reduce uncertainty. Demonstration projects and pilot schemes also play an important role by allowing communities to observe system performance and outcomes in real-world settings. Social acceptance is closely linked to regulatory credibility and governance transparency. When reuse guidelines are clearly defined and consistently enforced, public confidence in treated wastewater quality tends to increase. Conversely, past failures, pollution incidents, or weak regulatory oversight can undermine trust and lead to resistance, even when technical standards are met (Pathiranage *et al.*, 2024).

8.5.4 Engagement with smart technologies and digital systems

The integration of smart technologies into wastewater management, such as real-time monitoring, automated controls, and data-driven decision-support systems, introduces new social considerations. While these technologies offer potential benefits in terms of efficiency, early warning, and transparency, their acceptance depends on how stakeholders perceive data use, accountability, and control. Operators and regulators may require training to interpret data outputs and integrate them into routine decision-making. Without adequate support, advanced monitoring systems risk being underutilized or misinterpreted. At the community level, concerns related to data privacy, system reliability, and perceived loss of local control can influence acceptance of smart infrastructure (Nagpal *et al.*, 2024). Inclusive engagement strategies can help bridge these gaps by explaining how digital tools improve service quality and environmental protection. Public access to selected performance indicators, for example, can enhance transparency and accountability, strengthening trust between service providers and users. When smart technologies are introduced as supportive tools rather than surveillance mechanisms, they are more likely to gain acceptance (Dbouk *et al.*, 2026).

8.5.5 Equity, inclusion, and social sustainability

Social sustainability in wastewater management requires attention to equity and inclusion. Marginalized groups often bear a disproportionate burden of inadequate sanitation and environmental pollution, while having limited voice in decision-making processes. Inclusive stakeholder engagement seeks to address these imbalances by ensuring representation across gender, income levels, and social groups (Canova *et al.*, 2025). Targeted outreach, flexible engagement formats, and collaboration with local organizations can help include voices that are otherwise overlooked. Equity considerations are particularly important in tariff setting, service expansion, and reuse planning, where financial and social impacts may differ across user groups. When wastewater projects are perceived as fair and responsive to community needs, long-term acceptance and cooperation are more likely.

8.5.6 Linking social engagement to long-term system performance

Social dimensions and stakeholder engagement are closely linked to the long-term performance of wastewater systems. Projects that align technical solutions with social expectations tend to experience fewer conflicts, higher compliance, and more reliable operation. Engagement processes also provide feedback loops that allow institutions to adapt policies and practices over time. By integrating community participation, capacity building, and public engagement into governance frameworks, wastewater management can move beyond infrastructure provision toward socially embedded and resilient systems. Recognizing social factors as integral rather than peripheral elements of wastewater management strengthens the ability of policies and technologies to deliver sustainable outcomes across diverse contexts (Duque-Achipiz *et al.*, 2025; Monzón-Reyes *et al.*, 2025; Pathiranage *et al.*, 2024).

8.6 CONCLUSION

Sustainable wastewater management cannot be achieved through technological advancement alone; it depends fundamentally on the quality of governance and the strength of institutional arrangements that guide decision-making and implementation. Adaptive, transparent, and inclusive governance systems provide the conditions under which policies, technologies, and social priorities can be aligned in a meaningful and lasting manner. When governance frameworks are responsive to environmental change, demographic pressures, and evolving societal expectations, wastewater systems are better equipped to remain functional and relevant over time. Aligning policy objectives with technological capabilities requires a clear understanding of local institutional capacity and socio-economic conditions. Regulatory ambitions that exceed operational or financial realities risk poor compliance and system failure, while under-ambitious policies may limit innovation and environmental protection. Effective governance therefore lies in balancing environmental goals with practical considerations, ensuring that standards, financing mechanisms, and implementation strategies are mutually reinforcing. Institutional coordination and stakeholder engagement further strengthen the sustainability of wastewater systems by fostering shared responsibility and trust. Clear role definition, consistent enforcement, and inclusive participation help translate policy intent into operational practice. Ultimately, wastewater governance that integrates policy coherence, institutional capacity, and social acceptance offers a pathway toward resilient systems that protect public health, conserve resources, and support long-term environmental sustainability.

REFERENCES

Abreham Awono Y. and Adam D. B. (2025). Evaluation of operational challenges and technological performance of wastewater treatment plants in Addis Ababa. *Scientific Reports*, **15**(1), 33099, https://doi.org/10.1038/s41598-025-17945-8

Afego P., Aina I. V., Motohashi K., Narita D. and Thiam D. R. (2025). Governance of agricultural wastewater management in Japan and South Africa: a comparative institutional analysis. *Sustainable Water Resources Management*, **11**(6), 122, https://doi.org/10.1007/s40899-025-01297-2

Alexandre J., Kathumbi L., Javanbakht G. and Allport J. M. (2025). Effects of operation and maintenance costs on the financial sustainability of micro hydropower schemes. *Sustainable Energy Technologies and Assessments*, **83**, 104651, https://doi.org/10.1016/j.seta.2025.104651

Al-Rajhi A. M. H. and Abdelghany T. M. (2025). Innovative strategies for wastewater treatment: harnessing green technologies for sustainable resource recovery. A review. *Frontiers in Environmental Science*, **13**, 1696485. https://doi.org/10.3389/fenvs.2025.1696485

Altowayti W. A. H., Shahir S., Othman N., Eisa T. A. E., Yafooz W. M. S., Al-Dhaqm A., Soon C. Y., Yahya I. B., Rahim N., Abaker M. and Ali A. (2022). The role of conventional methods and artificial intelligence in the wastewater treatment: a comprehensive review. *Processes*, **10**(9), 1832, https://doi.org/10.3390/pr10091832

Canova J. T., Nahar S. and Moss T. W. (2025). From stigma to solution: sanitation and sustainable entrepreneurship in emerging economies. *Journal of Business Venturing Insights*, **23**, e00532, https://doi.org/10.1016/j.jbvi.2025.e00532

Dbouk H. M., Attieh L., Dbouk N. and Ezzeddine M. (2026). Wastewater treatment technologies and digital integration: a review of pathways toward circularity and sustainability. *Journal of Environmental Chemical Engineering*, **14**(2), 121718, https://doi.org/10.1016/j.jece.2026.121718

Duque-Achipiz N. A., Benitez-Campo N., Manyoma-Velásquez P., Curtis T. and Gómez-Borraz T. (2025). Community participation and technical design for wastewater management in Cassava Agro-Industry in Cauca, Colombia. *Water Practice and Technology*, **20**(5), 1237–1251, https://doi.org/10.2166/wpt.2025.060

Elliott M. A., Bakchan A., Maxcy-Brown J., D'Amato V. A., Hallahan D., White K. D., Stallman C. and Bradley S. (2023). Sustainable wastewater management for underserved communities using federal infrastructure funds: barriers, bottlenecks, and tradeoffs. *Water Security*, **20**, 100152, https://doi.org/10.1016/j.wasec.2023.100152

Ferrero G., Setty K., Rickert B., George S., Rinehold A., DeFrance J. and Bartram J. (2019). Capacity building and training approaches for water safety plans: a comprehensive literature review. *International Journal of Hygiene and Environmental Health*, **222**(4), 615–627, https://doi.org/10.1016/j.ijheh.2019.01.011

Gamache K. R. (2021). Critical infrastructure: water and wastewater systems sector. In: Encyclopedia of Security and Emergency Management, L. R. Shapiro and M.-H. Maras (eds), Springer International Publishing, Springer, Cham. pp. 171–181, https://doi.org/10.1007/978-3-319-70488-3_63

Halalsheh M. and Kassab G. (2018). Policy and the governance framework for wastewater irrigation: Jordanian experience. In: Safe Use of Wastewater in Agriculture: From

Concept to Implementation, H. Hettiarachchi and R. Ardakanian (eds), Springer International Publishing, Springer, Cham. pp. 75–99, https://doi.org/10.1007/978-3-319-74268-7_5

Hernández-Chover V., Bellver-Domingo Á., Castellet-Viciano L. and Hernández-Sancho F. (2025). Economies of scale and efficiency in the wastewater treatment sector: a decision tree approach. *Applied Sciences*, **15**(7), 3423, https://doi.org/10.3390/app15073423

Hughes J., Cowper-Heays K., Olesson E., Bell R. and Stroombergen A. (2021). Impacts and implications of climate change on wastewater systems: a New Zealand perspective. *Climate Risk Management*, **31**, 100262, https://doi.org/10.1016/j.crm.2020.100262

Jain H. (2025a). Data analytics enabled by the Internet of Things and artificial intelligence for the management of Earth's resources. In: Data Analytics and Artificial Intelligence for Earth Resource Management, D. Kumar, T. Tewary and S. Shekhar (eds), Elsevier, Amsterdam, Netherlands. Chapter 2, pp. 19–36, https://doi.org/10.1016/B978-0-443-23595-5.00002-4

Jain H. (2025b). Synergizing carbon reduction technologies and nature-based solutions for a climate-neutral future. *Mitigation and Adaptation Strategies for Global Change*, **30**(6), 42, https://doi.org/10.1007/s11027-025-10237-6

Jain H., Singh L., Shrivastava A., Paul M., Horo H. and Samuel H. S. (2026). MXene-enhanced polymer nanocomposite membranes for next-generation wastewater treatment and desalination. *Next Materials*, **10**, 101574, https://doi.org/10.1016/j.nxmate.2025.101574

Maliga I., Purwono S., Harini R. and Soetarto A. (2025). Critical indicators for determining sustainable domestic wastewater management for the achievement of SDG goal 6. *Discover Water*, **5**(1), 39, https://doi.org/10.1007/s43832-025-00235-9

Monzón-Reyes B. L., González-Moreno H. R., Month A. E. Á., Peralta Vega A. J., Ballut-Dajud G. and Sandoval Herazo L. C. (2025). Wastewater management strategies in rural communities using constructed wetlands: the role of community participation. *Earth*, **6**(2), 18, https://doi.org/10.3390/earth6020018

Nagpal M., Siddique M. A., Sharma K., Sharma N. and Mittal A. (2024). Optimizing wastewater treatment through artificial intelligence: recent advances and future prospects. *Water Science and Technology*, **90**(3), 731–757, https://doi.org/10.2166/wst.2024.259

Pathiranage W. B., Bray L., Jones K., Redwine N., Saralvarez J. and D'Alessio M. (2024). Perception and acceptance towards water reuse in the Southeast United States: a public survey. *Science of The Total Environment*, **908**, 168224, https://doi.org/10.1016/j.scitotenv.2023.168224

Piubello Orsini L., Leardini C., Danesi L., Guerrini A. and Frison N. (2023). Circular economy in the water and wastewater sector: tariff impact and financial performance of SMARTechs. *Utilities Policy*, **83**, 101593, https://doi.org/10.1016/j.jup.2023.101593

Rodrigues P. M., Pinto F. S. and Marques R. C. (2024). A framework for enabling conditions for wastewater reuse. *Sustainable Production and Consumption*, **46**, 355–366, https://doi.org/10.1016/j.spc.2024.02.022

Shamshad J. and Ur Rehman R. (2025). Innovative approaches to sustainable wastewater treatment: a comprehensive exploration of conventional and emerging technologies [10.1039/D4VA00136B]. *Environmental Science: Advances*, **4**(2), 189–222, https://doi.org/10.1039/D4VA00136B

Silva J. A. (2022). Implementation and integration of sustainability in the water industry: a systematic literature review. *Sustainability*, **14**(23), 15919, https://doi.org/10.3390/su142315919

Silva J. A. (2023). Wastewater treatment and reuse for sustainable water resources management: a systematic literature review. *Sustainability*, **15**(14), 10940, https://doi.org/10.3390/su151410940

Singh B. J., Chakraborty A. and Sehgal R. (2023). A systematic review of industrial wastewater management: evaluating challenges and enablers. *Journal of Environmental Management*, **348**, 119230, https://doi.org/10.1016/j.jenvman.2023.119230

Soo A. and Shon H. K. (2024). A nutrient circular economy framework for wastewater treatment plants. *Desalination*, **592**, 118090, https://doi.org/10.1016/j.desal.2024.118090

Tshililo F. P., Mutanga S. S., Gumbo T., Sikhwivhilu K. and Chikozho C. (2024). The influence of public perception on acceptance of potable water reclamation schemes. The case of Diepsloot township, Johannesburg. *Water Reuse*, **14**(1), 30–38, https://doi.org/10.2166/wrd.2024.108

Victor A. U., Singh L., Jain H. and Khan A. A. (2026). Ternary carbon quantum dot-based nanocomposites for sustainable water treatment: synergistic design, mechanisms, and challenges: a review. *Inorganic Chemistry Communications*, **186**, 116353, https://doi.org/10.1016/j.inoche.2026.116353

doi: 10.2166/9781789065404_0163

Chapter 9

Global case studies, emerging challenges, and future directions in smart wastewater management

Derya Ucbaş[1], Iyabo Christianah Oladipo[2], Sukhmandeep Kaur[3], Ditimoni Dutta[4], Shivani Jaswal[5], Harshita Jain[6] and Lovepreet Singh[7]

[1]*Department of Food Engineering, Eskişehir Osmangazi University, Eskişehir, Turkey. E-mail: derya.ucbas@ogu.edu.tr*
[2]*Department of Science Laboratory Technology (Microbiology Unit), Ladoke Akintola University of Technology, PMB 4000, Ogbomoso, Nigeria. E-mail: icoladipo@lautech.edu.ng*
[3]*Research Scholar, Department of Food Science and Technology, Punjab Agricultural University, Ludhiana, Punjab 141001, India. E-mail: sukmandeepkaur1999@gmail.com*
[4]*Department of Food Technology, Parul Institute of Applied Sciences, Parul University, Vadodara 391760, India. E-mail: ditimoni.dutta42805@paruluniversity.ac.in*
[5]*Department of Food Science and Technology, Dr. Y.S. Parmar University of Horticulture and Forestry, Solan 173230, India. E-mail: shivanijaswal0205@gmail.com*
[6]*Department of Climate Change and Environmental Sustainability, Amity Institute of Environment and Sustainability, Amity University Uttar Pradesh (AUUP), Noida, Uttar Pradesh 201301, India. E-mail: hjain@amity.edu*
[7]*Department of Chemical Engineering, Thapar Institute of Engineering and Technology, Patiala 147004, India. E-mail: lovepreet.singh@thapar.edu*
**Corresponding author: lovepreet.singh@thapar.edu*

ABSTRACT

The practical relevance of advanced wastewater management approaches becomes evident through their application in real-world settings. This chapter brings together selected case studies from different regions to illustrate how smart technologies, innovative treatment processes, and sustainable governance models are being implemented in diverse environmental and socio-economic contexts. The case studies highlight both successful outcomes and the practical constraints encountered during planning, operation, and long-term system management. Beyond documenting technological performance, the chapter examines how local conditions such as regulatory frameworks, institutional capacity, financial resources, and public acceptance shape project outcomes. Comparative insights are used to identify common challenges, including system scalability, operational reliability, and data integration, as well as strategies adopted to address them. The chapter concludes by looking ahead, discussing emerging trends and research directions that are likely to influence the future of wastewater management. These include increasing system decentralization, greater reliance on digital tools, and closer integration with energy and resource recovery systems. By connecting practical experience with forward-looking perspectives, this chapter provides a closing synthesis that links innovation with long-term sustainability goals.

Keywords: Case studies, smart wastewater systems, implementation challenges, future trends, sustainable water management, technology adoption

9.1 INTRODUCTION

Case studies play a crucial role in strengthening wastewater management practice because they reflect how treatment systems and management strategies perform under real operating conditions. While research studies and pilot-scale experiments are important for testing new ideas, they are usually conducted in controlled environments that do not fully represent the complexity of real communities, infrastructure limitations, or institutional settings. Case studies capture these complexities by documenting experiences from actual projects, making them an essential source of practical knowledge for both practitioners and decision-makers (Cairone *et al.*, 2024).

One of the key contributions of case studies is their ability to demonstrate how technologies function over time. Wastewater systems are long-term investments that must operate reliably under varying influent loads, seasonal changes, and evolving regulatory requirements. Case studies provide insights into system durability, maintenance needs, operational flexibility, and cost implications, offering a more complete picture than short-term performance data. They also reveal how unexpected challenges such as equipment failures, data management issues, or shortages of skilled personnel are addressed in practice (Borowska *et al.*, 2025).

Beyond technical performance, case studies highlight the importance of governance and institutional arrangements. Many wastewater projects succeed or fail not because of the technology itself, but due to factors such as fragmented responsibilities, limited financial resources, or weak regulatory enforcement. Real-world examples show how coordination between utilities, municipalities, regulators, and communities influences outcomes. They also illustrate the role of policy support, funding mechanisms, and capacity-building initiatives in sustaining wastewater systems over the long term (Reymond *et al.*, 2020).

Case studies are equally valuable in understanding social and economic dimensions. Public acceptance, user behavior, and community engagement often determine whether new treatment approaches or reuse schemes are successfully adopted. Through documented experiences, case studies show how communication strategies, stakeholder involvement, and transparency help build trust and encourage participation. They also reveal the consequences of neglecting these aspects, such as resistance to reuse projects or underutilization of infrastructure (Jain, 2025b; Shamshad & Ur Rehman, 2025).

In the context of smart wastewater management, case studies are particularly important because digital tools and automated systems rely heavily

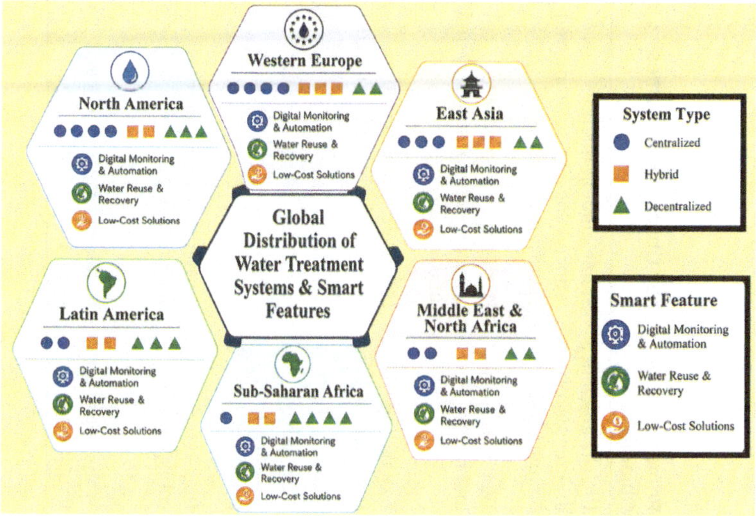

Figure 9.1 Global distribution of smart wastewater case studies across regions.

on continuous operation, data reliability, and user competence. Real-world examples provide evidence on how sensors, monitoring platforms, and decision-support systems perform over extended periods and under varying conditions. They also shed light on challenges related to data integration, cybersecurity, and system interoperability, which are difficult to assess through short-term trials (Hernández-Chover *et al.*, 2025; Jain, 2025a).

Importantly, case studies support learning across regions and contexts (Figure 9.1). By comparing experiences from different geographical and socio-economic settings, common patterns and recurring challenges can be identified. At the same time, case studies highlight the need for context-specific solutions, showing that approaches successful in one location may require adaptation elsewhere. This comparative learning helps guide technology transfer, policy formulation, and future research priorities. Case studies act as a bridge between innovation and implementation. They ground wastewater management decisions in real experience, support adaptive planning, and help align technological development with institutional capacity and social realities. By providing evidence of what works, what does not, and why, case studies contribute to more resilient, efficient, and sustainable wastewater management practices.

9.2 REGIONAL CASE STUDIES OF SMART WASTEWATER SYSTEMS

Smart wastewater management has taken different forms across regions, shaped by local priorities, levels of infrastructure development, regulatory strength, financial capacity, and social needs. Examining case studies from both developed and developing regions helps illustrate how similar goals such as improved treatment efficiency, water reuse, and environmental protection are pursued through very different implementation pathways (Jain *et al.*, 2026;

Table 9.1 Summary of regional smart wastewater case studies.

Region/Country	System Type	Key Smart Features	Scale of Implementation	Primary Objective	Key Observed Outcomes
Northern Europe (e.g., Denmark, Germany)	Centralized	Real-time sensors, automated aeration, digital twins	City-scale treatment plants	Energy efficiency and regulatory compliance	Stable effluent quality, reduced energy use, improved operational control
East Asia (e.g., Singapore, Japan)	Centralized with reuse integration	Advanced monitoring, reclaimed water control systems	National urban utilities	Water security and reuse	High public acceptance, reliable reuse for industry and landscaping
India	Decentralized/ Hybrid	Remote monitoring, low-cost sensors	Community and peri-urban scale	Expanding treatment access	Improved local sanitation coverage, reduced untreated discharge
Sub-Saharan Africa	Decentralized	Mobile-based monitoring, modular units	Small towns and settlements	Affordable treatment solutions	Improved system oversight despite limited staffing
Latin America	Centralized upgrades	Digital compliance reporting, SCADA systems	Medium-sized cities	Regulatory transparency	Better reporting, improved accountability
Middle East & North Africa	Centralized reuse-focused	Automated quality control	Agricultural reuse systems	Water reuse for irrigation	Enhanced reuse safety and farmer confidence

Singh *et al.*, 2023; Victor *et al.*, 2026). These regional experiences underline that there is no single model for smart wastewater systems; instead, solutions evolve in response to local conditions and constraints (Table 9.1).

9.2.1 Case studies from developed regions

In many developed countries, smart wastewater systems have emerged as an extension of well-established centralized infrastructure (Figure 9.2). Regions in Europe, North America, and parts of East Asia already possess extensive sewer networks and advanced treatment plants. In these contexts, 'smartness' is often introduced through digital upgrades rather than complete system redesigns. For example, several cities in Northern Europe have integrated real-time monitoring sensors into existing wastewater treatment plants to track flow rates, pollutant loads, and energy consumption. These systems allow operators to adjust aeration, chemical dosing, and sludge handling dynamically, leading to improved treatment efficiency and reduced energy use (Chavhan *et al.*, 2025).

A notable pathway in developed regions involves the use of advanced data analytics and automation. Utilities increasingly rely on digital twins, virtual models of treatment plants that mirror real-time operations. By comparing simulated outcomes with actual performance, operators can anticipate equipment failures, optimize maintenance schedules, and test operational changes without disrupting plant function. Case studies from countries such as Denmark and Germany show that these approaches have reduced operational costs and improved compliance with stringent discharge standards, particularly for nutrients such as nitrogen and phosphorus (Kaur *et al.*, 2025).

Another important trend in developed regions is the integration of wastewater systems with broader urban sustainability goals. In cities like Singapore and Tokyo, wastewater is no longer viewed solely as a waste stream but as a resource for water reuse, energy recovery, and nutrient recycling. Smart control systems are used to manage reclaimed water distribution for industrial use, urban landscaping, and groundwater recharge. These systems depend heavily on reliable monitoring and public trust, supported by strong regulatory frameworks and long-term investment planning (Narendra *et al.*, 2025).

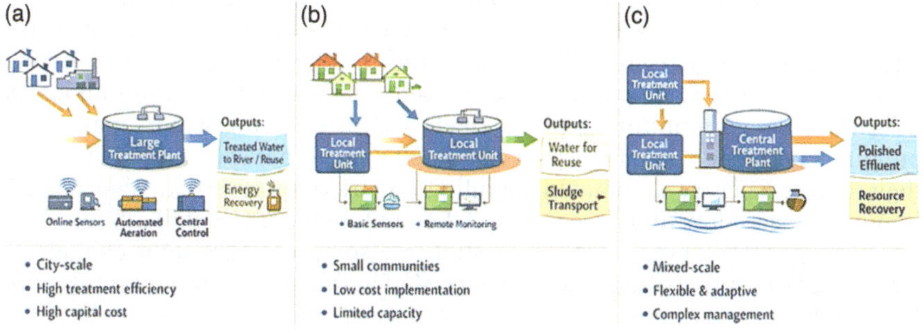

Figure 9.2 Typology of smart wastewater system configurations.

Despite these advantages, case studies from developed regions also highlight challenges. Legacy infrastructure can limit the pace of digital transformation, as older treatment plants may not be easily compatible with modern sensors and control systems. Data management and cybersecurity have emerged as new concerns, particularly as utilities become more reliant on cloud-based platforms. Moreover, high capital costs and the need for specialized technical expertise can slow adoption in smaller municipalities, even within otherwise well-resourced countries.

9.2.2 Case studies from developing regions

In developing regions, smart wastewater systems often follow very different pathways. Many cities in Asia, Africa, and Latin America face rapid urbanization, limited sewer coverage, and financial constraints. In these contexts, smart solutions are frequently linked with decentralization and modular system design rather than large centralized plants. Case studies from India, Indonesia, and parts of Sub-Saharan Africa show how small-scale treatment units equipped with basic sensors and remote monitoring are being deployed in peri-urban areas and informal settlements (Atofarati *et al.*, 2025).

These decentralized systems are designed to be flexible and cost-effective. For example, community-level treatment units using constructed wetlands or anaerobic digesters are increasingly paired with low-cost sensors that monitor parameters such as pH, turbidity, and flow. Data are often transmitted via mobile networks to centralized dashboards, allowing operators or local authorities to oversee multiple units with limited staff. Such approaches have improved system reliability and reduced downtime, especially in areas where on-site technical expertise is scarce (Ventura *et al.*, 2024).

Another common pathway in developing regions involves public–private partnerships and donor-supported initiatives. Case studies from Latin America illustrate how international funding agencies and local governments collaborate to introduce smart monitoring systems in wastewater treatment plants serving medium-sized cities. These projects often focus on improving regulatory compliance and transparency, using digital tools to generate reliable performance data that can be shared with regulators and the public. While these initiatives have improved accountability, their long-term sustainability depends on continued financial and institutional support (Kyambade *et al.*, 2025).

Water reuse is also a major driver of smart wastewater systems in water-stressed developing regions. In parts of the Middle East and North Africa, treated wastewater is increasingly reused for agriculture and landscape irrigation. Smart monitoring plays a key role in ensuring water quality and building confidence among users. Case studies show that even relatively simple digital tools, such as automated sampling and alert systems, can significantly improve reuse safety and acceptance (Ukoha-Onuoha & Horsfall, 2026).

However, developing-region case studies also reveal persistent challenges. Financial limitations often restrict the level of automation and data integration that can be achieved. Power supply interruptions, limited internet connectivity, and shortages of skilled personnel can undermine system performance. In some cases, digital tools are installed but not fully used due to lack of training or unclear institutional responsibility. These experiences highlight that technology alone is insufficient; capacity building and governance reform are equally important (Khan *et al.*, 2025).

9.2.3 Comparative insights and diverse pathways

When examined together, regional case studies reveal both contrasts and shared lessons. Developed regions tend to focus on optimizing existing systems through advanced analytics, automation, and integration with energy and resource recovery. Developing regions, in contrast, prioritize access, affordability, and flexibility, often combining simple smart tools with decentralized treatment approaches. Yet, both pathways reflect a common shift toward data-informed decision-making and proactive system management. A key insight from comparative analysis is that smart wastewater systems evolve incrementally. Even in technologically advanced settings, digital transformation is often phased, starting with basic monitoring before progressing to predictive control and integrated management platforms. Similarly, in developing regions, modest digital interventions can deliver significant benefits when aligned with local needs and capacities (Wang *et al.*, 2025).

Another shared lesson is the importance of institutional context. Successful case studies consistently demonstrate clear roles and responsibilities, supportive policies, and stakeholder engagement. Where governance is weak or fragmented, even well-designed smart systems struggle to deliver sustained benefits. Conversely, strong institutional frameworks can compensate for technological limitations by ensuring consistent operation and maintenance. Regional case studies underscore that smart wastewater management is not defined by the sophistication of technology alone. Instead, it is shaped by how technologies are selected, adapted, and embedded within local environmental, social, and institutional contexts. By learning from diverse regional experiences, planners and policymakers can design wastewater systems that are not only technically sound but also practical, inclusive, and resilient across different development settings (Opoku Mensah *et al.*, 2026).

9.3 COMPARATIVE ANALYSIS OF IMPLEMENTATION OUTCOMES

A comparative analysis of smart wastewater management initiatives across regions provides valuable insight into how different systems perform in practice and why outcomes vary. While many projects share similar objectives such as improving treatment efficiency, reducing environmental impacts, and enhancing resource recovery their results are shaped by local technical conditions, operational capacities, and broader institutional and social contexts. By examining performance outcomes alongside operational experiences, this section highlights key lessons that can inform future wastewater planning and management (Table 9.2).

9.3.1 Performance assessment across systems

Performance assessment in smart wastewater systems goes beyond conventional indicators such as pollutant removal efficiency or compliance with discharge standards. Case studies show that system performance must be evaluated across multiple dimensions, including reliability, energy use, operational flexibility,

Table 9.2 Comparative assessment of implementation outcomes.

Performance Aspect	Developed Regions	Developing Regions	Key Influencing Factors	Common Limitations
Treatment efficiency	High and stable	Moderate to improving	Process control, operator skills	Influent variability
Energy performance	Optimized through automation	Limited optimization	Energy cost, technology access	Power interruptions
Operational reliability	High	Variable	Maintenance planning, spare parts	Delayed repairs
Data utilization	Advanced analytics	Basic monitoring	Institutional capacity	Underuse of collected data
Resource recovery	Integrated and planned	Emerging	Policy support, financing	Market access for recovered products
Long-term sustainability	Strong	Uneven	Financial stability	Dependence on external funding

and long-term cost effectiveness. In many developed regions, smart systems have demonstrated clear improvements in process control, leading to more stable effluent quality and reduced chemical and energy consumption. Real-time monitoring and automated controls allow operators to respond quickly to fluctuations in influent load, seasonal changes, or equipment malfunction, thereby maintaining consistent performance (Capodaglio, 2025) (Figure 9.3).

In developing regions, performance outcomes are more mixed but often significant in relative terms. Even basic monitoring and remote supervision have led to noticeable reductions in system downtime and untreated discharges. Decentralized treatment units equipped with simple digital tools have improved local treatment coverage and enabled timely maintenance interventions. However, performance gains are sometimes constrained by external factors such as intermittent power supply, limited spare parts, or inconsistent inflow patterns. These case studies suggest that performance improvements should be judged in relation to baseline conditions rather than against idealized benchmarks (Martí-Herrero *et al.*, 2025).

Across both contexts, resource recovery performance has emerged as an increasingly important outcome. Smart systems that integrate energy recovery, nutrient recycling, or water reuse tend to perform better when digital monitoring is aligned with clear operational objectives. For example, biogas production from anaerobic digestion is more stable when feedstock quality and process conditions are continuously tracked. Similarly, water reuse schemes benefit from automated quality monitoring that ensures safety and builds user confidence. Comparative evidence indicates that systems designed with resource recovery in

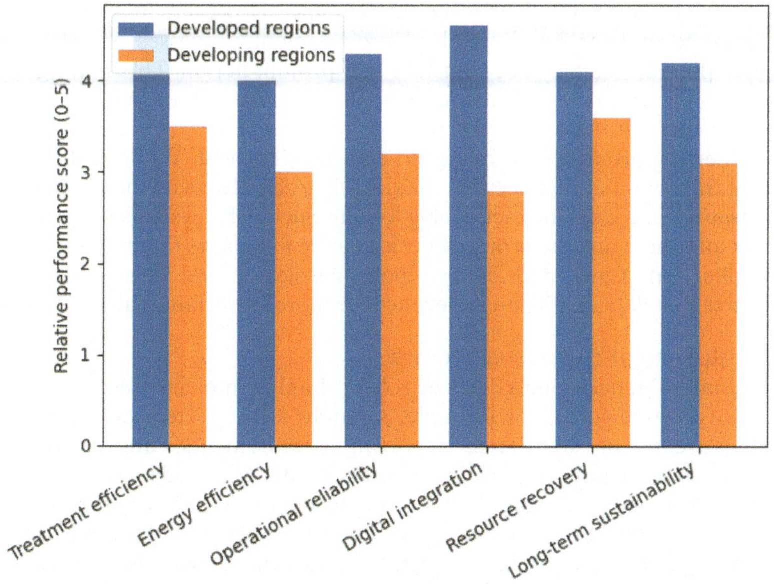

Figure 9.3 Comparative performance outcomes of smart wastewater systems.

mind from the outset tend to achieve more consistent results than those where recovery is added later as an afterthought (Das & Mishra, 2025).

9.3.2 Operational lessons from real-world practice

Operational experience is one of the most valuable contributions of comparative analysis. Across regions, case studies consistently show that technology performance is closely linked to operator capacity and day-to-day management practices. In advanced systems, highly automated processes can reduce manual workload, but they also require skilled personnel who can interpret data, troubleshoot issues, and make informed decisions. Where training and institutional support are strong, smart systems enhance operational efficiency; where they are lacking, automation may lead to system underuse or operational errors (Deepu & Ravi, 2023).

In developing contexts, operational lessons often revolve around simplicity and adaptability. Systems that rely on robust, low-maintenance technologies combined with basic digital monitoring tend to perform more reliably than highly complex installations. Case studies reveal that over-designed systems frequently face operational difficulties when local capacity does not match technical demands. As a result, many successful projects adopt a phased approach, gradually increasing the level of automation as experience and confidence grow.

Maintenance practices also play a critical role in shaping outcomes. Smart systems generate large volumes of data, but this information is only useful if it leads to timely maintenance actions. Comparative evidence shows that preventive maintenance supported by predictive analytics can significantly reduce breakdowns and extend equipment life. However, in settings where maintenance budgets are limited or procurement processes are slow, even accurate predictions may not translate into prompt action. This highlights the need to align digital tools with realistic maintenance planning and financial processes (Benhanifia *et al.*, 2025).

Another important lesson relates to data management. While real-time data improve transparency and decision-making, they can also overwhelm operators if not presented in a clear and actionable format. Successful systems prioritize user-friendly interfaces and focus on key performance indicators rather than excessive data collection. Case studies from both developed and developing regions emphasize that data quality and relevance are more important than data quantity.

9.3.3 Influence of contextual factors

Contextual influences play a decisive role in shaping implementation outcomes. Regulatory frameworks, for example, strongly affect system performance by defining treatment standards, reporting requirements, and enforcement mechanisms. In regions with clear and consistently enforced regulations, smart systems are more likely to be used proactively to maintain compliance and optimize performance. In contrast, weak or inconsistent regulation can reduce incentives for effective operation, even when advanced technology is available.

Financial context is another major influence. High upfront investment costs can limit the adoption of smart technologies, particularly in resource-constrained settings. Comparative case studies show that projects with stable long-term financing and clear cost-recovery mechanisms tend to perform better over time. Conversely, systems heavily dependent on short-term grants or external funding often struggle to sustain operations once initial support ends. This is especially evident in cases where digital platforms require ongoing subscription fees or regular software updates (Dafri & Al-Qaruty, 2023).

Institutional arrangements also shape outcomes by determining roles, responsibilities, and coordination among stakeholders. Systems managed by well-defined utilities with clear accountability structures generally perform more consistently than those involving fragmented governance. In some cases, overlapping responsibilities between municipal departments or between public and private actors have led to delays in decision-making and reduced system effectiveness. Successful examples highlight the value of clear institutional mandates and strong inter-agency coordination.

Social and cultural factors further influence implementation outcomes. Public perception of wastewater reuse, trust in authorities, and willingness to adopt new practices all affect system success. Comparative analysis shows that early and continuous stakeholder engagement improves acceptance and reduces resistance, particularly in reuse-oriented projects. Where community concerns are ignored, even technically sound systems may face opposition or underutilization (Akpan *et al.*, 2020).

9.3.4 Cross-cutting insights and implications

Taken together, comparative analysis reveals that smart wastewater systems succeed when performance objectives, operational capacity, and contextual conditions are aligned. Advanced technology alone does not guarantee better outcomes; it must be supported by skilled operators, appropriate governance, and sustainable financing. At the same time, even modest digital interventions can deliver meaningful benefits when they are well matched to local needs and constraints (Benhanifia *et al.*, 2025; Dafri & Al-Qaruty, 2023; Khan *et al.*, 2025).

A key implication for future implementation is the importance of flexibility. Systems that allow gradual scaling, modular upgrades, and adaptive management are better able to respond to changing conditions. Comparative evidence also underscores the need for integrated planning that considers technical, institutional, and social dimensions from the outset. The comparative analysis of implementation outcomes demonstrates that learning from diverse experiences is essential for improving wastewater management practice. By understanding how performance, operations, and context interact, practitioners and policymakers can design smarter systems that are resilient, effective, and suited to the realities of their operating environments (Das & Mishra, 2025; Khan *et al.*, 2025; Opoku Mensah *et al.*, 2026; Wang *et al.*, 2025).

9.4 COMMON CHALLENGES ACROSS WASTEWATER MANAGEMENT PROJECTS

Despite advances in treatment technologies and management approaches, wastewater projects across the world continue to face a set of recurring challenges. These challenges are not limited to a particular region or level of development; rather, they emerge in different forms depending on local context. Understanding these shared barriers is essential for designing systems that are not only technically sound but also practical, durable, and socially acceptable (Table 9.3).

Table 9.3 Common challenges across wastewater management projects.

Challenge Category	Description	Typical Causes	Impact on Project Outcomes
Technical	Equipment malfunction and system instability	Aging infrastructure, sensor fouling	Reduced treatment efficiency
Financial	Insufficient funds for O&M	Low tariffs, limited subsidies	Premature system failure
Institutional	Fragmented governance	Overlapping agency roles	Delayed decision-making
Institutional	Limited skilled workforce	Inadequate training	Underuse of smart systems
Social	Resistance to reuse	Low awareness, trust deficits	Underutilized treated water
Social	Unequal service coverage	Informal settlements	Social inequity and health risks

9.4.1 Technical challenges

Technical challenges remain one of the most visible obstacles in wastewater management projects. Many systems struggle with variability in influent quantity and quality, driven by seasonal changes, industrial discharges, or rapid urban growth. Treatment technologies that perform well under steady conditions often face difficulties when inflow patterns fluctuate, leading to reduced efficiency or operational instability. In smart wastewater systems, the reliability of sensors and monitoring equipment becomes an additional concern, particularly in harsh environments where fouling, corrosion, or power fluctuations are common (Liang & Yue, 2021).

Another technical issue relates to system complexity. Advanced treatment processes and automated controls can improve performance, but they also increase dependence on specialized equipment and skilled operators. In several case studies, systems have underperformed because local operators were not adequately trained to manage complex control platforms or interpret monitoring data. Compatibility between old infrastructure and new digital tools is also a frequent challenge, as many existing plants were not designed to accommodate modern sensors or automation systems.

Maintenance is a further technical barrier. Wastewater infrastructure requires regular upkeep, yet maintenance planning is often reactive rather than preventive. Delays in repairing equipment, lack of spare parts, and limited access to technical support can lead to prolonged system downtime. These issues are particularly pronounced in decentralized or remote systems, where logistical constraints make routine maintenance more difficult (Lawson *et al.*, 2024).

9.4.2 Financial challenge

Financial limitations are a major constraint across wastewater management projects. High capital costs for treatment plants, monitoring systems, and digital platforms can delay project initiation or result in scaled-down designs. Even when initial funding is secured, long-term operational and maintenance costs are frequently underestimated. Many systems struggle to remain functional once external funding or project-based support ends.

Cost recovery mechanisms also present challenges. In many regions, wastewater tariffs do not fully reflect the true cost of treatment, making utilities dependent on government subsidies. This situation can restrict investment in upgrades, staff training, and preventive maintenance. Smart wastewater technologies, while offering efficiency gains, often involve recurring expenses related to software licensing, data management, and equipment replacement, which can strain limited budgets.

Uncertainty in funding streams further complicates planning. Projects reliant on short-term grants or donor assistance may face interruptions if funding priorities change. Financial challenges are often closely linked to institutional capacity, as weak financial management systems can hinder effective budgeting and expenditure tracking (Banerjee *et al.*, 2022).

9.4.3 Institutional challenges

Institutional barriers play a critical role in shaping project outcomes. Fragmented governance structures, where responsibilities for planning, operation, regulation, and enforcement are divided among multiple agencies, often lead to coordination problems. In such cases, decision-making can be slow, and accountability becomes unclear, affecting system performance and responsiveness.

Limited institutional capacity is another common challenge. Many utilities operate with insufficient staffing, high turnover, or limited access to specialized expertise. This affects not only day-to-day operations but also the ability to adopt and sustain new technologies. In smart wastewater projects, lack of institutional readiness can result in digital tools being underused or abandoned altogether.

Policy and regulatory inconsistencies also pose barriers. Outdated standards, weak enforcement, or frequent policy changes can create uncertainty for project developers and operators. Without clear regulatory signals, investments in advanced treatment or reuse systems may be delayed. Institutional challenges are often compounded by inadequate coordination between national, regional, and local authorities (Villarín & Merel, 2020).

9.4.4 Social challenges

Social barriers are sometimes less visible but equally influential. Public perception of wastewater treatment and reuse can strongly affect project acceptance. In many regions, negative attitudes toward treated wastewater limit the uptake of reuse schemes, even when water quality meets safety standards. Lack of awareness and limited engagement with communities during planning stages often contribute to resistance or mistrust.

Equity concerns also arise, particularly in rapidly urbanizing areas. Wastewater infrastructure may not reach informal settlements or low-income communities, leading to uneven service provision. Projects that fail to consider social inclusion risk reinforcing existing inequalities and may face opposition or non-cooperation from affected groups.

Behavioral factors further influence system performance. Improper waste disposal, illegal connections, or industrial discharges can disrupt treatment processes and damage infrastructure. Addressing these issues requires not only technical solutions but also education, enforcement, and community involvement (Valipour & Singh, 2016).

9.4.5 Interconnected nature of challenges

Importantly, these challenges rarely occur in isolation. Technical problems are often linked to financial constraints, while institutional weaknesses can exacerbate social resistance. Case studies show that projects are more resilient when challenges are addressed in an integrated manner, combining appropriate technology with sustainable financing, strong institutions, and active stakeholder engagement. Common challenges across wastewater management projects highlight the need for balanced and context-sensitive approaches. Recognizing and addressing technical, financial, institutional, and social barriers early in the project lifecycle can improve the likelihood of long-term success and ensure that wastewater systems deliver intended environmental and public health benefits.

9.5 EMERGING TRENDS AND RESEARCH PRIORITIES

Wastewater management is undergoing a gradual but meaningful shift, driven by growing pressure on water resources, climate uncertainty, urban expansion, and rising expectations for environmental performance. Instead of focusing only on end-of-pipe treatment, current thinking increasingly views wastewater systems as flexible service networks that must respond to local needs, recover resources, and remain functional under stress. Emerging trends and research priorities reflect this shift, emphasizing approaches that are practical, adaptive, and aligned with long-term sustainability goals (Table 9.4).

9.5.1 Decentralized and hybrid treatment systems

One of the most visible trends is the growing interest in decentralized and hybrid wastewater systems. Traditional centralized treatment plants have clear advantages in terms of economies of scale, but they are often expensive to expand and difficult to adapt to rapid urban growth or dispersed settlements. Decentralized systems, including community-scale plants and on-site treatment units, offer greater flexibility and faster deployment. They are particularly relevant in peri-urban areas, small towns, and regions where sewer networks are incomplete or absent (Figure 9.4).

Research is increasingly focused on improving the reliability and performance of decentralized systems under varying conditions. This includes studies on modular design, low-energy treatment processes, and simplified operation and maintenance requirements. Hybrid models that combine centralized and decentralized elements are also gaining attention, as they allow cities to extend coverage while reducing pressure on existing infrastructure. Future research needs to address long-term monitoring, governance arrangements, and cost-sharing mechanisms to ensure that decentralized systems remain effective beyond the pilot stage (Lema, 2025).

Table 9.4 Emerging trends and priority research areas in smart wastewater management.

Emerging Trend	Key Focus Areas	Current Gaps	Priority Research Needs
Decentralized systems	Modular design, low-energy processes	Long-term performance data	Lifecycle assessment, governance models
Digital integration	Real-time monitoring, automation	Data overload, cybersecurity	User-oriented interfaces, secure data systems
Circular economy linkages	Nutrient and energy recovery	Economic feasibility	Market integration, policy alignment
Resilience planning	Climate adaptation, risk management	Limited stress-testing	Climate-responsive system design
Hybrid systems	Central–decentral integration	Coordination complexity	Planning and optimization frameworks

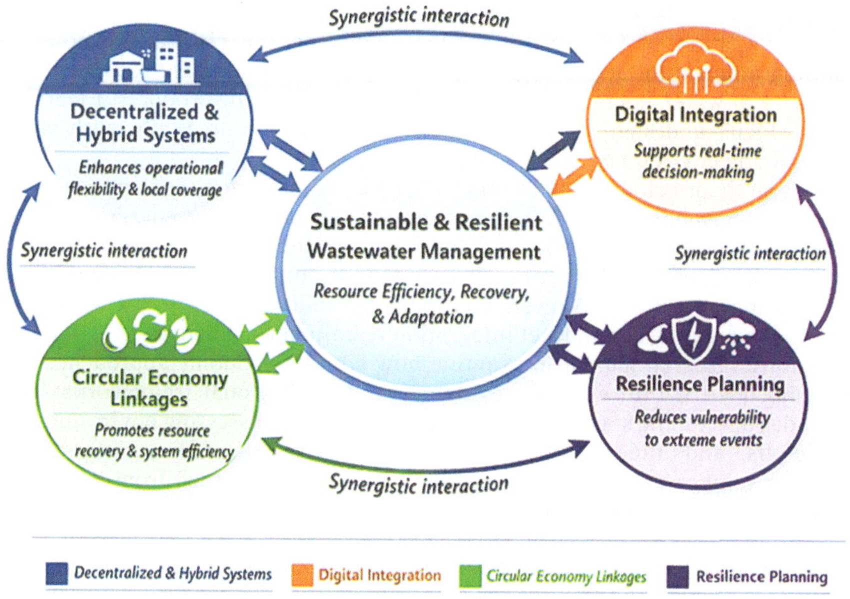

Figure 9.4 Conceptual framework linking emerging trends in smart wastewater management.

9.5.2 Digital integration and data-driven management

Digital tools are becoming central to modern wastewater management, but their role is evolving from basic monitoring to integrated decision support. Sensors, remote monitoring, and automation are now being combined with data analytics to improve process control, predict failures, and support preventive maintenance. These developments allow operators to move from reactive management toward more informed and proactive approaches (Expósito & Díez Cebollero, 2025).

Current research priorities include improving data quality, interoperability, and usability. Many systems collect large volumes of data but struggle to translate information into clear operational actions. There is growing interest in user-centered system design, where digital platforms are tailored to the needs and skills of operators rather than purely technical capabilities. Cybersecurity and data governance have also emerged as important research areas, particularly as wastewater utilities become more connected and reliant on digital infrastructure.

Another emerging trend is the use of digital tools to improve transparency and regulatory compliance. Real-time reporting and open data platforms can strengthen accountability and build trust among regulators and the public. Research is needed to understand how these tools influence institutional behavior and long-term performance (Essamlali *et al.*, 2024).

9.5.3 Circular economy and resource recovery linkages

The integration of wastewater management into the circular economy is gaining momentum across regions. Wastewater is increasingly recognized as a source of water, energy, nutrients, and other valuable by-products. Emerging trends focus on designing systems that recover these resources efficiently while minimizing environmental impacts.

Research priorities include improving technologies for nutrient recovery, such as phosphorus and nitrogen capture, and enhancing energy recovery through anaerobic digestion and biogas utilization. There is also growing interest in recovering heat, organic compounds, and even rare elements from wastewater streams. However, technical feasibility alone is not sufficient; economic viability and market integration remain key challenges.

Future research must also examine how circular economy approaches can be embedded within existing regulatory and institutional frameworks. This includes developing standards for recovered products, assessing environmental trade-offs, and understanding user acceptance. Case studies suggest that circular solutions are more successful when they are planned from the outset rather than added as secondary objectives (Sewenet *et al.*, 2026).

9.5.4 Resilience and climate-responsive planning

Resilience planning has emerged as a critical priority in wastewater management, particularly in the face of climate change and increasing frequency of extreme events. Floods, droughts, heatwaves, and power disruptions pose significant risks to wastewater infrastructure and service continuity. Emerging trends emphasize designing systems that can withstand, adapt to, and recover from such stresses.

Research in this area focuses on risk assessment, flexible system design, and emergency response planning. This includes exploring nature-based solutions, such as wetlands and green infrastructure, that can complement engineered systems while providing additional environmental benefits. There is also growing interest in integrating wastewater planning with broader urban resilience strategies, including energy systems, stormwater management, and land-use planning.

Digital tools play an important role in resilience planning by enabling early warning, real-time response, and post-event analysis. Future research needs to examine how these tools can be effectively integrated into routine operations without increasing system complexity or cost (Banerjee *et al.*, 2022; Essamlali *et al.*, 2024; Expósito & Díez Cebollero, 2025).

9.5.5 Linking research with practice

Across all emerging trends, a key research priority is strengthening the link between innovation and everyday practice. Many promising ideas fail to move beyond pilot projects because they do not adequately account for institutional capacity, financing, or user behavior. Future research should therefore place greater emphasis on long-term performance, scalability, and adaptability. Emerging trends in wastewater management point toward systems that are more decentralized, digitally enabled, resource-oriented, and resilient. Research priorities must support this transition by focusing not only on technological advancement but also on governance, capacity building, and social acceptance. Together, these efforts can help shape wastewater systems that are better prepared to meet future environmental and societal challenges (Lema, 2025; Sewenet *et al.*, 2026; Villarín & Merel, 2020).

9.6 CONCLUSION

Experiences from wastewater management projects across different regions clearly show that successful outcomes depend on more than the choice of technology. While technical innovation plays an important role, its effectiveness is strongly shaped by local environmental conditions, institutional structures, financial capacity, and social acceptance. Global case studies demonstrate that solutions must be adapted to the specific context in which they are implemented, rather than being transferred as fixed models. Approaches that work well in one region may require significant modification elsewhere to account for differences in infrastructure, governance, and community needs. Another key lesson is the value of adaptive management. Wastewater systems operate in dynamic settings where influent characteristics, regulatory requirements, and resource demands change over time. Projects that allow for flexibility, learning, and gradual improvement tend to be more resilient and reliable in the long term. The use of monitoring, feedback, and periodic evaluation enables operators and decision-makers to respond to challenges as they arise and adjust system design or operation accordingly. Future progress in wastewater management will depend on how well technological advances are aligned with institutional capacity and policy frameworks. Digital tools, automation, and resource recovery technologies can deliver meaningful benefits only when supported by skilled personnel, clear roles and responsibilities, and consistent regulatory oversight. Equally important is the availability of stable financing mechanisms that support not just initial investment but also long-term operation and maintenance. Finally, long-term sustainability objectives must remain central to wastewater planning and management. This includes protecting public health and ecosystems, improving water security, and promoting efficient use of resources. By grounding innovation in real-world experience and aligning it with governance, policy, and social realities, wastewater management systems can evolve to meet future challenges in a balanced and sustainable manner.

REFERENCES

Akpan V. E., Omole D. O. and Bassey D. E. (2020). Assessing the public perceptions of treated wastewater reuse: opportunities and implications for urban communities in developing countries. *Heliyon*, **6**(10), e05246, https://doi.org/10.1016/j.heliyon.2020.e05246

Atofarati E. O., Adogbeji V. O. and Enweremadu C. C. (2025). Sustainable smart waste management solutions for rapidly urbanizing African Cities. *Utilities Policy*, **95**, 101961, https://doi.org/10.1016/j.jup.2025.101961

Banerjee A., Sarkar A. and Ghosal S. (2022). Challenges and strategies of water supply and wastewater management in coastal urban and semi-urban areas. In: Current

Directions in Water Scarcity Research, A. L. Srivastav, S. Madhav, A. K. Bhardwaj and E. Valsami-Jones (eds), Elsevier, Amsterdam, Netherlands. Vol. 6, pp. 425–446, https://doi.org/10.1016/B978-0-323-91838-1.00011-7

Benhanifia A., Cheikh Z. B., Oliveira P. M., Valente A. and Lima J. (2025). Systematic review of predictive maintenance practices in the manufacturing sector. *Intelligent Systems with Applications*, **26**, 200501, https://doi.org/10.1016/j.iswa.2025.200501

Borowska W., Zubrowska-Sudol M. and Doskocz N. (2025). Full-scale wastewater treatment plant monitoring: insights from batch assays on nitrogen removal. *Journal of Water Process Engineering*, **77**, 108641, https://doi.org/10.1016/j.jwpe.2025.108641

Cairone S., Hasan S. W., Choo K.-H., Lekkas D. F., Fortunato L., Zorpas A. A., Korshin G., Zarra T., Belgiorno V and Naddeo V. (2024). Revolutionizing wastewater treatment toward circular economy and carbon neutrality goals: pioneering sustainable and efficient solutions for automation and advanced process control with smart and cutting-edge technologies. *Journal of Water Process Engineering*, **63**, 105486, https://doi.org/10.1016/j.jwpe.2024.105486

Capodaglio A. G. (2025). Energy audits and energy efficiency of urban wastewater systems, following UWWTP directive 2024/3019. *Water*, **17**(14), 2049, https://doi.org/10.3390/w17142049

Chavhan N., Bhattad R., Khot S., Patil S., Pawar A., Pawar T. and Gawli P. (2025). APAH: an autonomous IoT driven real-time monitoring system for industrial wastewater. *Digital Chemical Engineering*, **14**, 100217, https://doi.org/10.1016/j.dche.2025.100217

Dafri W. and Al-Qaruty R. (2023). Challenges and opportunities to enhance digital financial transformation in crisis management. *Social Sciences & Humanities Open*, **8**(1), 100662, https://doi.org/10.1016/j.ssaho.2023.100662

Das A. and Mishra S. (2025). Smart and sustainable water management: policy, technology, and innovation pathways for resource resilience. *Green Technology, Resilience, and Sustainability*, **5**(1), 5, https://doi.org/10.1007/s44173-025-00022-8

Deepu T. S. and Ravi V. (2023). A review of literature on implementation and operational dimensions of supply chain digitalization: framework development and future research directions. *International Journal of Information Management Data Insights*, **3**(1), 100156, https://doi.org/10.1016/j.jjimei.2023.100156

Essamlali I., Nhaila H. and El Khaili M. (2024). Advances in machine learning and IoT for water quality monitoring: a comprehensive review. *Heliyon*, **10**(6), e27920, https://doi.org/10.1016/j.heliyon.2024.e27920

Expósito A. and Díez Cebollero E. (2025). How the digital revolution is reshaping water management and policy: a focus on Spain. *Utilities Policy*, **96**, 102020, https://doi.org/10.1016/j.jup.2025.102020

Hernández-Chover V., Bellver-Domingo Á., Castellet-Viciano L. and Hernández-Sancho F. (2025). Economies of scale and efficiency in the wastewater treatment sector: a decision tree approach. *Applied Sciences*, **15**(7), 3423, https://doi.org/10.3390/app15073423

Jain H. (2025a). Data analytics enabled by the Internet of Things and artificial intelligence for the management of Earth's resources. In: Data Analytics and Artificial Intelligence for Earth Resource Management, D. Kumar, T. Tewary and S. Shekhar (eds), Elsevier, Amsterdam, Netherlands. pp. 19–36, https://doi.org/10.1016/B978-0-443-23595-5.00002-4

Jain H. (2025b). Synergizing carbon reduction technologies and nature-based solutions for a climate-neutral future. *Mitigation and Adaptation Strategies for Global Change*, **30**(6), 42, https://doi.org/10.1007/s11027-025-10237-6

Jain H., Singh L., Shrivastava A., Paul M., Horo H. and Samuel H. S. (2026). MXene-enhanced polymer nanocomposite membranes for next-generation wastewater treatment and desalination. *Next Materials*, **10**, 101574, https://doi.org/10.1016/j.nxmate.2025.101574

Kaur P., Szlachta M., Xu J., Vepsäläinen J., Lappalainen R., Hussin F. and Aroua M. K. (2025). Advancements and feasibility of synergistic approaches in phosphorus recovery from wastewater: a critical review. *Journal of Environmental Management*, **386**, 125786, https://doi.org/10.1016/j.jenvman.2025.125786

Khan M. I., Yasmeen T., Khan M., Hadi N. U., Asif M., Farooq M. and Al-Ghamdi S. G. (2025). Integrating industry 4.0 for enhanced sustainability: pathways and prospects. *Sustainable Production and Consumption*, **54**, 149–189, https://doi.org/10.1016/j.spc.2024.12.012

Kyambade M., Mwesigwa R., Alinda K., Tumwine S. and Lwanga F. (2025). Drivers of sustainability performance in health and sanitation projects: a systematic literature review. *Social Sciences & Humanities Open*, **12**, 102046, https://doi.org/10.1016/j.ssaho.2025.102046

Lawson E., Robbie J., Sloan W. T. and Amezaga J. (2024). Challenges and opportunities for decentralised water technologies in Scotland. *Water Policy*, **26**(2), 189–205, https://doi.org/10.2166/wp.2024.194

Lema M. W. (2025). Wastewater crisis in East African cities: challenges and emerging opportunities. *Discover Environment*, **3**(1), 18, https://doi.org/10.1007/s44274-025-00206-w

Liang X. and Yue X. (2021). Challenges facing the management of wastewater treatment systems in Chinese rural areas. *Water Science and Technology*, **84**(6), 1518–1526, https://doi.org/10.2166/wst.2021.332

Martí-Herrero J., Erazo I. and Jiménez-Paute R. (2025). Low-cost digesters and wetlands for decentralized wastewater treatment: performance and GHG emissions in the Ecuadorian Amazon. *Science of the Total Environment*, **1002**, 180646, https://doi.org/10.1016/j.scitotenv.2025.180646

Narendra K., Manjushree M., Adur A. J., Raajasubramanian D., Srinivasan S. and Murali R. (2025). Self-healing water systems: the future of smart, adaptive, and regenerative water networks. *Cleaner Water*, **4**, 100112, https://doi.org/10.1016/j.clwat.2025.100112

Opoku Mensah S., Jacobs B. and Cunningham R. (2026). The role of institutions in enhancing climate change adaptation by smallholder farmers in the Talensi district of Ghana. *Environmental Development*, **57**, 101341, https://doi.org/10.1016/j.envdev.2025.101341

Reymond P., Chandragiri R. and Ulrich L. (2020). Governance arrangements for the scaling up of small-scale wastewater treatment and reuse systems–lessons from India [Original Research]. *Frontiers in Environmental Science*, **8**, 2020, https://doi.org/10.3389/fenvs.2020.00072

Sewenet A. D., Boulaksil Y. and Pisano P. (2026). Circular economy, circularity, and sustainability: a systematic review and conceptual framework. *Cleaner Environmental Systems*, **20**, 100405, https://doi.org/10.1016/j.cesys.2026.100405

Shamshad J. and Ur Rehman R. (2025). Innovative approaches to sustainable wastewater treatment: a comprehensive exploration of conventional and emerging technologies. *Environmental Science: Advances*, **4**(2), 189–222, https://doi.org/10.1039/D4VA00136B

Singh B. J., Chakraborty A. and Sehgal R. (2023). A systematic review of industrial wastewater management: evaluating challenges and enablers. *Journal of Environmental Management*, **348**, 119230, https://doi.org/10.1016/j.jenvman.2023.119230

Ukoha-Onuoha E. and Horsfall I. T. (2026). The evolution of wastewater reuse for irrigation: lessons learned and future directions. *Next Sustainability*, **7**, 100253, https://doi.org/10.1016/j.nxsust.2026.100253

Valipour M. and Singh V. P. (2016). Global experiences on wastewater irrigation: challenges and prospects. In: Balanced Urban Development: Options and Strategies for Liveable Cities, B. Maheshwari, B. Thoradeniya and V. P. Singh (eds), Springer International Publishing, Cham. pp. 289–327, https://doi.org/10.1007/978-3-319-28112-4_18

Ventura J.-R. S., Tulipan J. U., Banawa A., Umali K. D. C. and Villanueva J. A. L. (2024). Advancements and challenges in decentralized wastewater treatment: a comprehensive review. *Desalination and Water Treatment*, **320**, 100830, https://doi.org/10.1016/j.dwt.2024.100830

Victor A. U., Singh L., Jain H. and Khan A. A. (2026). Ternary carbon quantum dot-based nanocomposites for sustainable water treatment: synergistic design, mechanisms, and challenges: a review. *Inorganic Chemistry Communications*, **186**, 116353, https://doi.org/10.1016/j.inoche.2026.116353

Villarín M. C. and Merel S. (2020). Paradigm shifts and current challenges in wastewater management. *Journal of Hazardous Materials*, **390**, 122139, https://doi.org/10.1016/j.jhazmat.2020.122139

Wang Q., Li Y. and Li R. (2025). Integrating artificial intelligence in energy transition: a comprehensive review. *Energy Strategy Reviews*, **57**, 101600, https://doi.org/10.1016/j.esr.2024.101600

doi: 10.2166/9781789065404_0183

Chapter 10

Integrated remote sensing and GIS approach for solid waste disposal site suitability mapping

Maya Kumari[1], Anamika Shrivastava[2] and Diksha[3]*

[1]Amity School of Natural Resources & Sustainable Development, Amity University, Sector 125, Noida, Uttar Pradesh, India
[2]Amity Institute of Environmental Sciences, Amity University, Sector 125, Noida, Uttar Pradesh, India
[3]Division of Environment Science, ICAR-Indian Agricultural Research Institute, IARI, Pusa Campus, New Delhi 110012
**Corresponding author: mkumar10@amity.edu*

ABSTRACT

Solid waste disposal is becoming a main global problem. Due to growing human activity, solid waste is developing extreme harm to the environment and human health. Damage is due to the illegal dumping of urban waste in undesirable places. For this reason, local municipal and central governments are moving ahead to assemble engineered solid waste landfills in suitable regions. A geographic information system (GIS) is a tool that can be utilized by engineers and urban planners to discover excellent reasonable sites for disposal. This observes attachment GIS and multi-criteria evaluation to develop a suitability version based on previous research of solid waste disposal site selection. Data was collected, and locations inside the capital region of Gautam Budh Nagar Uttar Pradesh, India were prioritized primarily based on the suitability model used in this study. The model evolved via this study use raster layers (land-use category, distance to railways and roads, slope, and drainage layer) classified to conduct suitability analysis. Suitable locations were then converted to polygons with the purpose of comparing their size and land-use type. Due to the extreme population of India, available land is restricted. This study utilized various techniques to become aware of available land suitable for solid waste disposal.

Keywords: Land-use land cover, urbanization, industrialization, solid waste management, environment

10.1 INTRODUCTION

Solid waste disposal is an environmental problem associated with urbanization and industrialization (Haripavan & Dey, 2023). Solid waste disposal is becoming a major global issue. Waste material is discharged from most parts of everyday human activity, resulting in adverse effects on human health and our surroundings (Fraternali *et al.*, 2024; Mussa & Suryabhagavan, 2019). Solid waste can have serious negative effects on human health and the environment. Solid waste refers to the unwanted or unusable materials generated from human activities in housing, industrial, or enterprise areas (Fraternali *et al.*, 2024; Hailu, 2019). It can be categorized based on fabric, consisting of plastic, paper, glass, metal, and organic matter (Abdel-Shafy & Mansour, 2018), or based on hazard potential, including radioactive, flammable, infectious, poisonous, or non-toxic materials. Additionally, categories may pertain to the source of the waste, such as organization, home, commercial enterprise, institutional or production, and demolition (Tadesse, 2004). Urban growth is one of the primary drivers for robust waste disposal (Okot *et al.*, 2019).

Solid waste management can be defined as the process associated with control of the generation, storage, collection, transport or transfer, processing, and disposal of stable waste materials in a way that effectively addresses public health, environmental, economic, aesthetic, and engineering concerns (Khan & Samadder, 2014; Saketa *et al.*, 2023). Within its scope, good waste management includes planning, administrative, economic, engineering, and legal functions. Solutions may also involve complex interdisciplinary relationships among fields such as public health, urban and local planning, political science, geography, sociology, economics, communication and conservation, demography, engineering, and materials sciences (Perdue *et al.*, 2003).

Solid waste management practices can vary between residential and enterprise generators, city and rural areas, and developed and developing countries (Aluko & Igwe, 2017; Bilgilioglu *et al.*, 2022; Sanjay *et al.*, 2018). The management of non-hazardous waste in metropolitan areas is usually the responsibility of neighborhood authorities. On the other hand, the control of hazardous waste materials is usually the responsibility of the waste generator, regulated by neighborhood, national, or perhaps international authorities.

It is critically important to identify appropriate areas for solid waste disposal. However, land is a limited resource, and its availability is increasingly constrained by rising land costs. If the land is chosen through a fair and systematic process, there is less likely to be opposition to the disposal site. Site selection for long-term disposal needs must therefore be done in a systematic way that does not compromise the local ecology. A centralized waste management approach may not be suitable for large cities or towns; therefore, greater emphasis should be placed on decentralized waste disposal systems which are often more effective and sustainable in the long term.

Sanitary disposal sites represent one of the most cost-effective methods of solid waste disposal for urban areas in developing countries (Gbanie *et al.*, 2013; Goel *et al.*, 2017; Vergara & Tchobanoglous, 2012). Recently, because of the growing need to resolve ecological issues in cities, solid waste management in low-income countries has received considerable attention, with a shift toward landfill site design aimed at enhancing environmental protection. One of principal issues is managing effective sites for waste disposal. Leachate from landfill or disposal sites typically contains significant levels of pollutants such as ammonia, nitrate, chloride, and heavy metals. These pollutants can infiltrate groundwater, damage water resources, or pose significant risks to human health. Landfill and waste site selection is a complex, multi-criteria, and often challenging process that requires the evaluation of numerous technical, economic and social factors. Environmental issues are especially critical, as waste sites can adversely affect the biophysical surroundings and ecosystems (Aderoju *et al.*, 2018; Al-Hanbali *et al.*, 2022; Ersoy & Bulut, 2009; Farooqi *et al.*, 2021).

Economic issues also have to be taken into consideration in landfill siting, including the costs of construction, operation, and maintenance of the facility. Social and cultural opposition to any landfill site can be a major impediment to the successful siting of waste disposal centers.

Solid waste management remains a significant global environmental challenge, as population growth and rising consumption generate increasing amounts of waste (Irhoumah *et al.*, 2014). Solid waste has become an inevitable by-product of economic development and rapid urbanization, with the challenges being particularly acute in less developed countries. Poor waste management, with rubbish dumped in water bodies or on open land without adequate treatment, has resulted in severe environmental pollution and associated health risks (Alexakis & Sarris, 2014).

Geographic information systems (GIS) and remote sensing are computerized systems that can be used to support effective planning and sustainable management of waste processes (Mallick, 2021). Site choice or suitability assessment is a type of analysis utilized in GIS to determine the most appropriate location for a given purpose. Potential sites considered in suitability evaluation may include areas for a new clinic, retail outlet or school, among many others (Chang *et al.*, 2008). When doing site analysis, users need to define the numerous requirements against which the GIS software evaluates the most suitable locations. Site selection analysis can be performed with either vector or raster data, but one of the most commonly used approaches uses raster information (Gautam *et al.*, 2020).

GIS undoubtedly presents enhanced information management, higher-quality analysis and improves performance. The use of GIS reduces time and cost, and enhances accuracy (Akbari *et al.*, 2008). Since the planning and decision-making processes are highly dependent on spatial data, GIS plays a vital role in solid waste management site selection (Das, 2020). Some of the key planning choices where GIS is essential are making route plans for vehicle transportation, finding the most appropriate locations for transfer stations, locating new landfill locations and monitoring all facilities. GIS is a tool that does not just decrease the time and cost of the location selection, but also offers a digital database for future analysis of the locations (Demesouka *et al.*, 2014; Kontos *et al.*, 2005; Rezaeisabzevar *et al.*, 2020; Sumathi *et al.*, 2008).

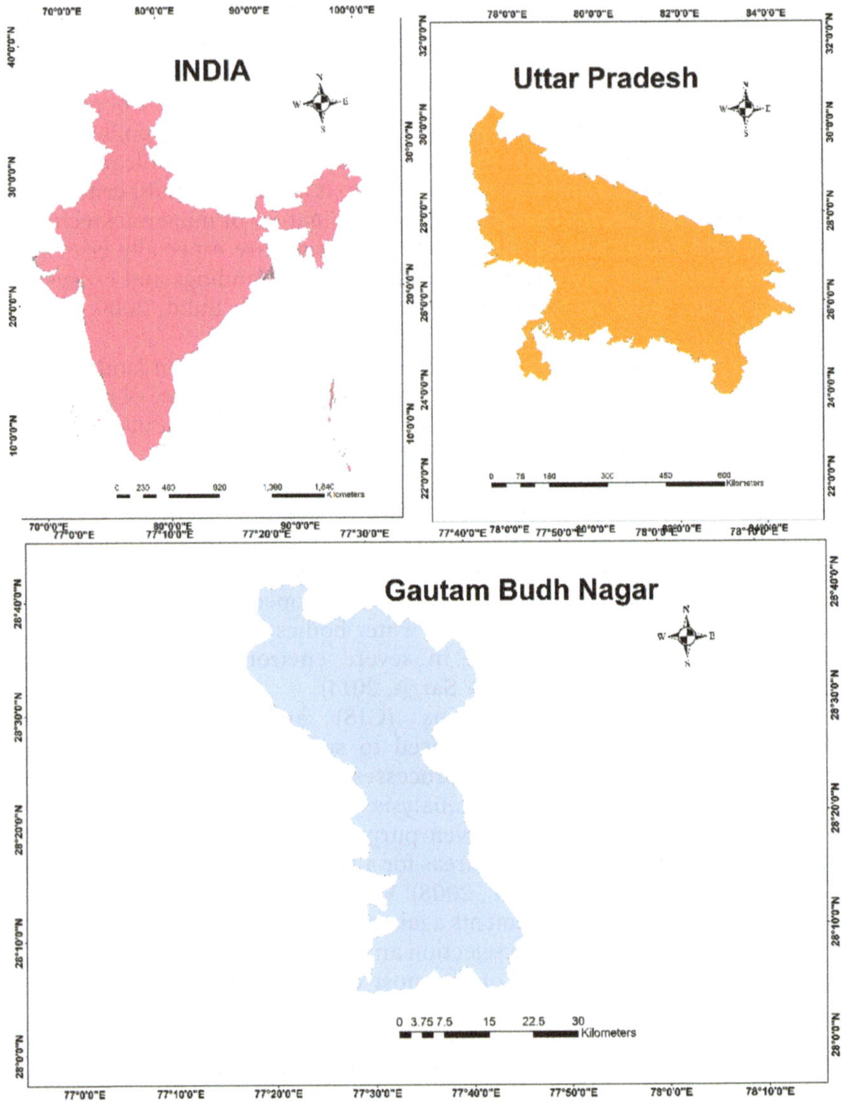

Figure 10.1 Geographical location of the study area.

Hence, GIS has become an indispensable device. However, the accuracy of the statistics significantly impacts the results (Ajibade *et al.*, 2019; Aksoy & San, 2019; Zeleke & Wedajo, 2025).

The issue of good waste disposal arrangements is addressed with effective control and regulation, along with waste reduction, reuse or recycling, and proper disposal practices (Chang *et al.*, 2008; Gbanie *et al.*, 2013; Gorsevski *et al.*, 2012; Santhosh & Sivakumar Babu, 2018). However, in the developing countries, municipal waste processes are often either inefficient or at a very basic level, and so the waste generated has become a significant threat to the environment. It has been observed that there is often a tendency to dispose of waste indiscriminately and in an uncoordinated manner, leading to environmental degradation and serious adverse effects on ecology and human health (Aksoy & San, 2019; Şener *et al.*, 2006). Numerous studies have demonstrated the effective application of GIS in waste disposal site selection and related analysis (Alkaradaghi *et al.*, 2019; Khan & Samadder, 2015; Şener *et al.*, 2011).

10.2 STUDY AREA

Gautam Budh Nagar is a residential district of Uttar Pradesh state in northern India, it is located at 28.3383°N, 77.6078°E. The total area of Gautam Budh Nagar is 1442 km^2 with its headquarters at the commercial town of Greater Noida, which has three Tehsils or four developmental blocks, viz., Bisrakh, Dadri, and Dankaur and Jewar. According to the 2011 census, the entire population of the district was 1 674 714. The district is known for its rapid urbanization and industrial development, with numerous multinational companies setting up their offices and factories in the area. It is also home to several prestigious educational institutions, making it an educational hub. Gautam Budh Nagar offers a high standard of living with modern infrastructure, well-planned residential complexes, and a variety of recreational facilities. The district's proximity to the national capital, Delhi, further enhances its appeal, attracting people from different parts of the country for work and settlement. The district's strategic location near Delhi has also led to the development of a robust transportation network, including highways and a metro line, making it easy for residents to commute to and from the capital city (Figure 10.1).

10.3 DATA AND SOFTWARE USED

10.3.1 Data used

Landsat 8 Operational Land Image (OLI) was downloaded for the year 2019 (April) to create Gautam Budh Nagar's land-use/landcover (LULC) map, and vector data were also extracted. ASTER DEM data was downloaded and used to create a slope and drainage map.

10.3.2 Software used

ArcGIS and Erdas IMAGINE software applications were utilized in creating and improving digital GIS statistics.

10.4 METHODOLOGY

The methodology is a logical part of the study to guide systematic investigation.

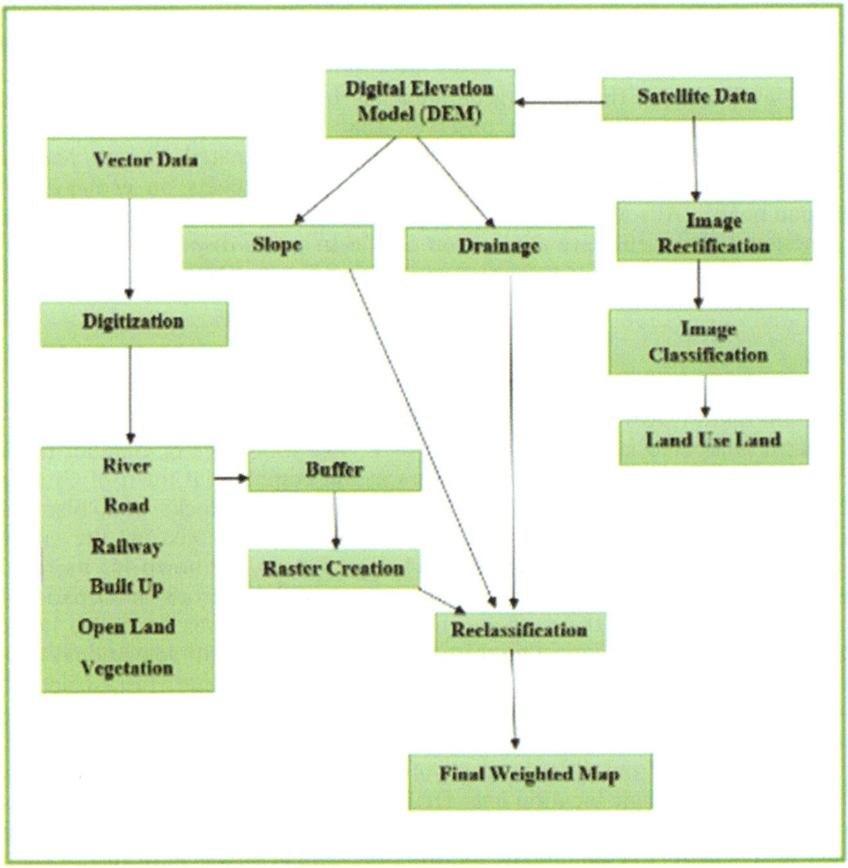

Figure 10.2 Flow chart of the methodology.

10.4.1 Data pre-processing

A multispectral image was created using the layer stack tool, stacked band 2 (blue), band 3 (green), band 4 (red), band 5 (NIR), band 6 (SWIR), and band 7 (SWIR) from the image Landsat 8 OLI. After stacking the layers, the subset of the study area was created.

10.4.2 Image classification

The land-use and land cover (LULC) map was created with the help of ArcGIS software and divided into six categories: built-up, vegetation, open land, water bodies (e.g., river, lake), road, and railway.

10.4.3 Buffering

Several ring buffers were created around the land-use and land cover maps, and these buffers were classified into four categories: 500 m, 1000 m, 1500 m, and 2000 m.

10.4.4 Reclassification and weighted mapping

The study used a weighted overlay analysis technique to identify a suitable solid waste site in the study region. The reclassification of the layers was divided into the 1, 2, 3, and 4 ranking system, whereby 1–unsuitable, 2–less suitable, 3–moderately suitable, 4–highly suitable. Then the satellite image was categorized according to different land uses to determine the land-use layers. These layers were interpolated in the Arc. New numerical values were given to specific classes for analysis. Eventually, most of these grid areas were mixed up in the weighted overlay analysis tool. Areas with lowest weights were identified as least suitable, while areas with the highest weights were identified as the most suitable places with minimal environmental impact (Figure 10.2).

10.5 RESULTS AND DISCUSSION

To minimize the negative environmental impacts of solid waste disposal sites, it is essential to follow established guidelines and criteria during site selection. Such sites should be located at a considerable distance from densely populated areas to protect public health. Furthermore, to ensure that waste disposal sites are environmentally sound and pose minimal risk, they must be carefully planned and comply with all relevant suitability criteria identified in the literature.

10.5.1 LULC classification

The LULC must be identified to decide which areas are more suitable for a landfill site. This is a basic map that makes it easy to create multiple thematic map layers that are required for the overlap analysis. The study area comprises different types of LULC classes, including built-up areas, open land, rivers, vegetation, roads and railway networks: all of which are shown in Figure 10.3.

10.5.1.1 Built-up areas

Landfills should not be placed too close to urban locations with a dense population. The study area was divided into three categories: areas with high density were considered as not suitable, areas with medium density were considered as less suitable, and areas with low density were considered as most suitable, as shown in Figure 10.5a. Landfill sites must not be located within 500 m of built-up areas.

Settlement areas are given high preference for reasons of public health and safety, as they represent concentrations of the human population. Increasing the distance from settlements also allows for expansion at a suitable stage in the selection of the location (Collins *et al.*, 2001).

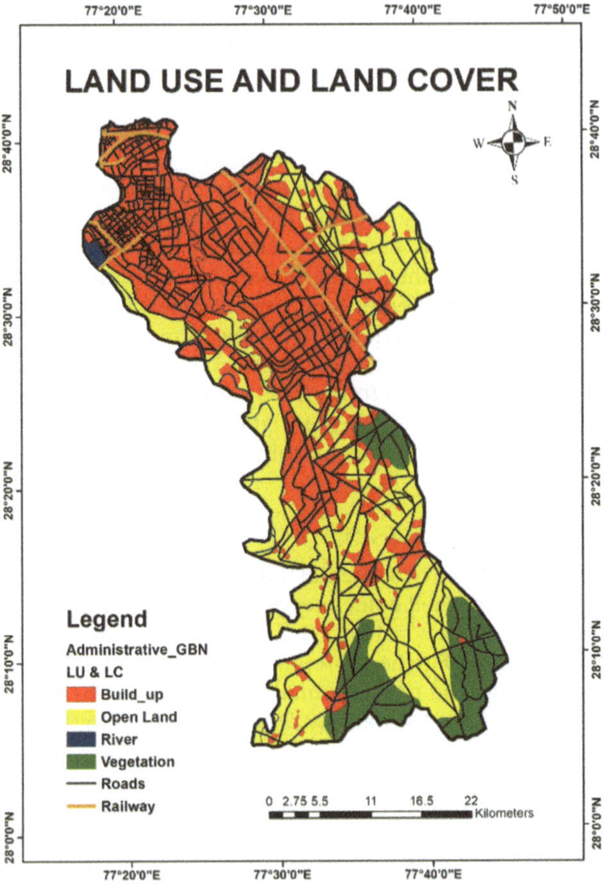

Figure 10.3 LULC map.

10.5.1.2 Vegetation

Agricultural land is a land cover used for agricultural purposes, particularly for growing crops that produce food for humans. Therefore, any land which might be used for agricultural purposes needs to be preserved and guarded from pollution. So, it is vital to keep solid waste disposal site/landfills away from agricultural land (Figure 10.4).

The green cover of the study area was divided into four permeability zones. The areas with high permeability are considered as unsuitable for landfill sites and are scored 1. This is because it would be possible for leachate infiltration to occur there, contaminating both the ground and surface of the neighboring areas (Babalola & Busu, 2011; Yal & Akgün, 2013). Areas with a low permeability rate were given much higher preference and are considered suitable for sanitary landfill sites, with scores of 4 (see Figure 10.5d).

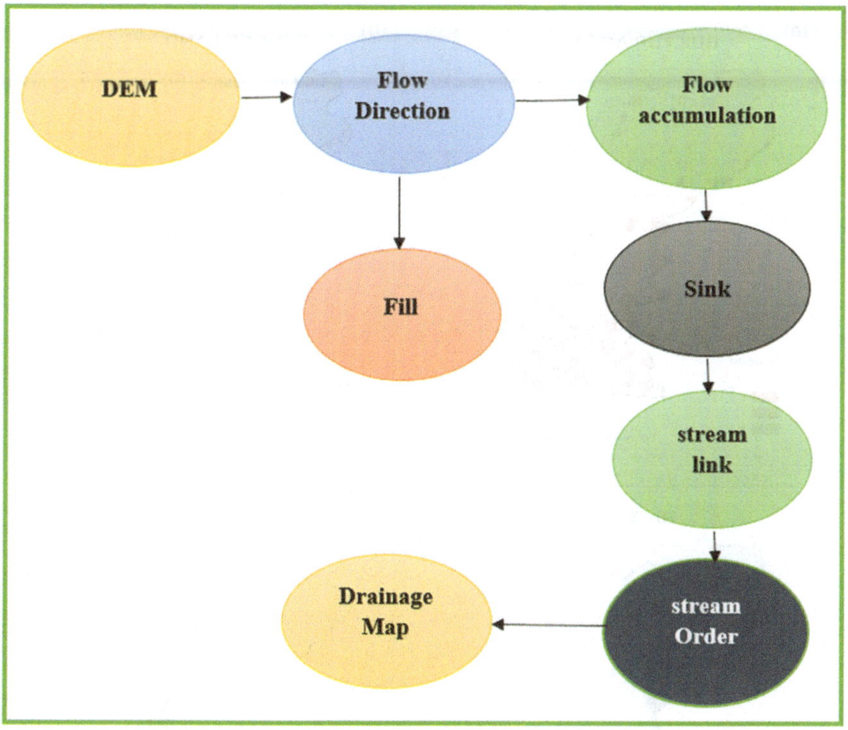

Figure 10.4 The drainage map workflow.

10.5.1.3 Open land

Open land areas are most suitable for a landfill site. The open land class must be identified so that one can determine which regions are more appropriate for landfill. Open land areas were classified with scores of 1 or 2 in the index of land-use suitability, as shown in Figure 10.5b. Open land is suitable because it is far away from built-up areas, vegetation, and river basins (Al-Hanbali *et al.*, 2011; Fatoyinbo *et al.*, 2020).

10.5.1.4 River

Sanitary landfills cannot be constructed adjacent to water bodies such as rivers and lakes due to the adverse environmental impacts which can occur; a water body can be contaminated by leachate generated from the landfill site area. Thus, a buffer zone of 500, 1000, 1500, 2000, and 8000 m was created for each of the rivers in the study area. Based on these buffer zones, we classified the river into four classes: (1) unsuitable, (2) less suitable, (3) moderately suitable, and (4) highly suitable, as shown in Figure 10.5c, and the value is assigned according to the criteria given for the suitability. Landfill sites close to the river pose a risk of water contamination and associated health issues.

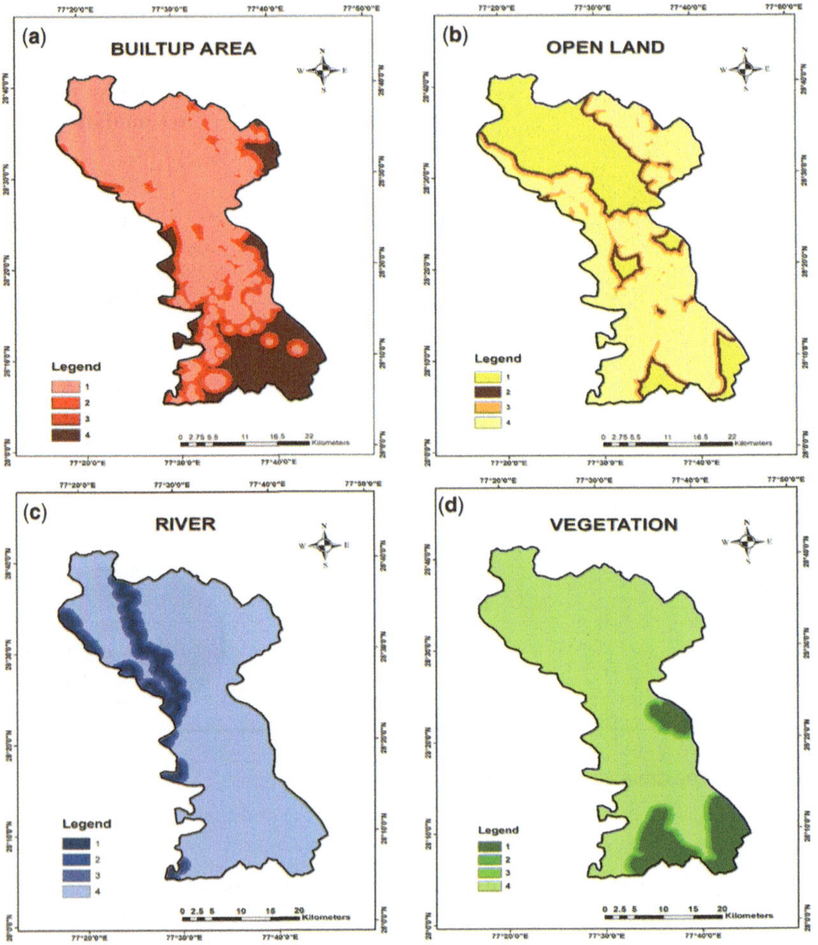

Figure 10.5 Layers of the LULC map. Reclassified maps (built-up (a), open land (b), river (c), vegetation (d)).

10.5.1.5 Road and railway

The road and railway network of the study area was extracted from the satellite OLS data. The road data includes information on highways, major roads, and their connections. This data is necessary for selecting landfill sites: the accessibility of the site must be considered, especially for vehicles used for the collection and disposal of waste to landfills (Balew *et al.*, 2022; Saketa *et al.*, 2023).

High scores were given to areas near the road. So, to identify which roads fit the criteria we performed a buffer of 500, 1000, 1500, and 2000 m around each road that is obtained from the data for the study area. According to it, road and railway zones were classified into unsuitable, less suitable, moderately suitable, and highly suitable. In terms of the economic context, landfills should not be located too far from the road, nor too close. The site should be a minimum of 1 km from the road, so according to this criteria, the results of the reclassified area are shown in Figure 10.6a and b.

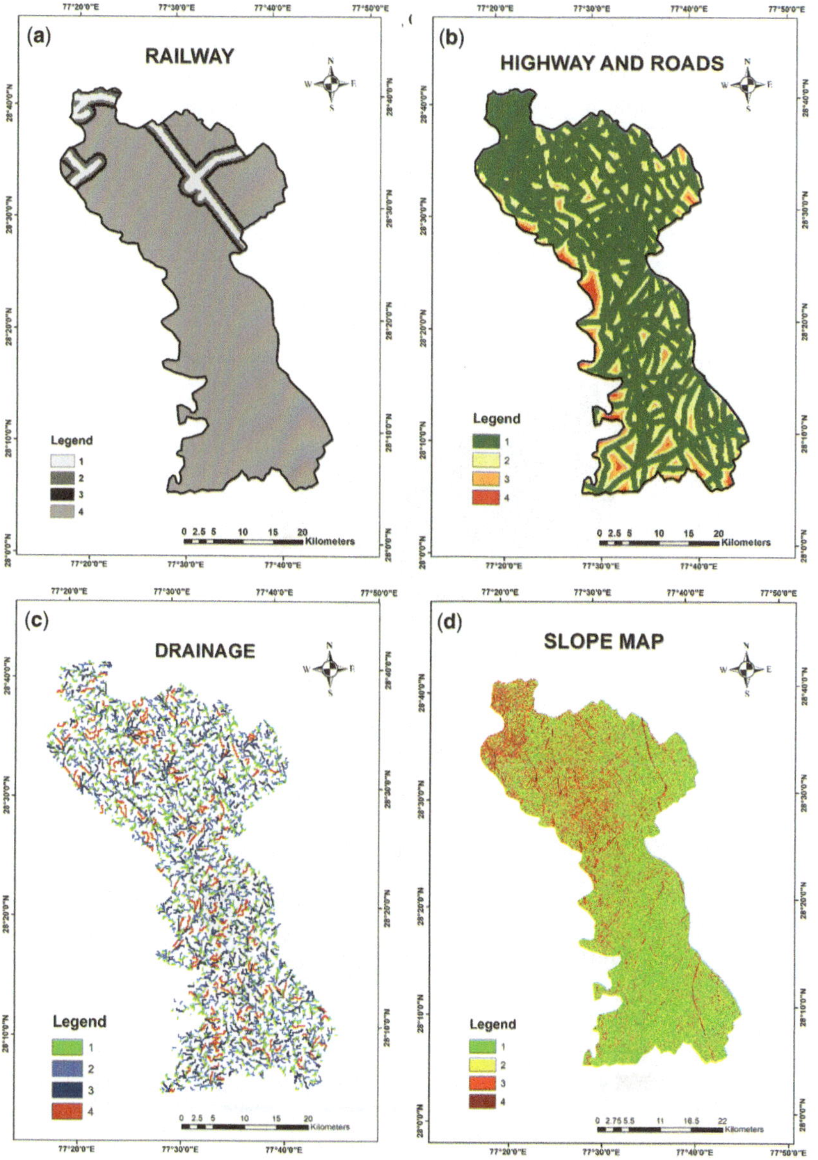

Figure 10.6 Reclassified maps (railway (a), highway and roads (b), drainage (c), slope (d)).

Table 10.1 Area covered by the different LULC classes.

Feature	Area/Length
Built-up area	683.327191 sq.km
Vegetation	142.235203 sq.km
River	12.397293 sq.km
Open land	547.717871 sq.km
Road	1676.983319 km
Railway	148.327467 km

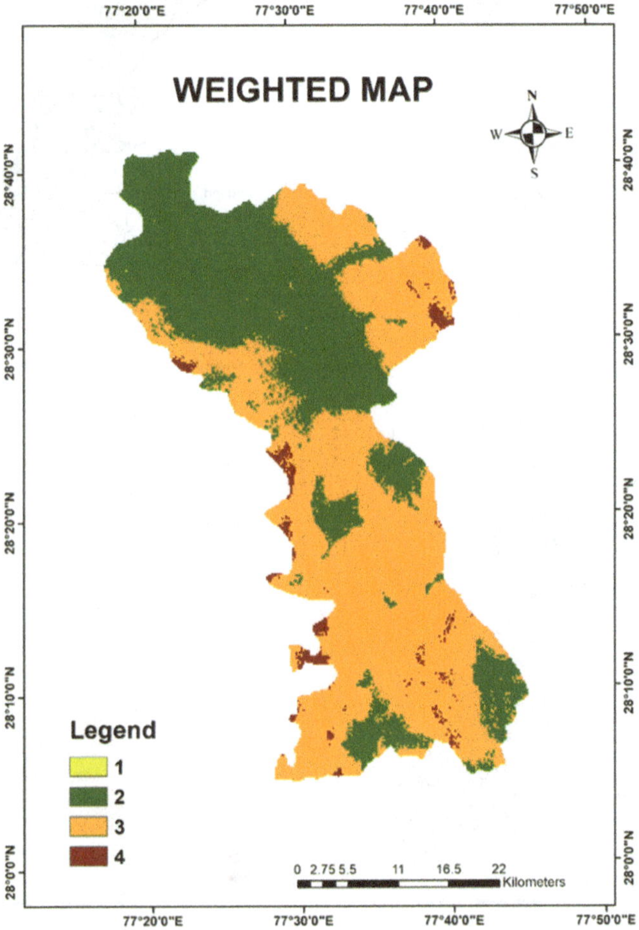

Figure 10.7 Weighted map.

The areas covered by the different LULC classes in April 2019 are shown in Table 10.1. According to the study, the built-up has an area of 683.327191 sq. km, vegetation has an area of 142.235203 sq.km, river has an area of 12.397293 sq.km, open land has an area of 547.717871 sq.km, road has an area of 1676.983319 km, and railway has an area of 148.327467 km.

Table 10.2 Weight of suitable solid waste disposal site selection factors.

Factor	Class	Value	Level of Suitability	Influence
Built-up	0–500	1	Unsuitable	
	500–1000	2	Less suitable	
	1000–1500	3	Moderately suitable	
	1500–11 000	4	Highly suitable	
Vegetation	0–500	1	Unsuitable	
	500–1000	2	Less suitable	
	1000–1500	3	Moderately suitable	
	1500–45 000	4	Highly suitable	
Open land	0–500	1	Unsuitable	
	500–1000	2	Less suitable	
	1000–1500	3	Moderately suitable	
	1500–14 500	4	Highly suitable	
River	0–500	1	Unsuitable	
	500–1000	2	Less suitable	
	1000–1500	3	Moderately suitable	
	1500–25 500	4	Highly suitable	
Road	0–500	1	Unsuitable	
	500–1000	2	Less suitable	
	1000–1500	3	Moderately suitable	
	1500–3000	4	Highly suitable	
Railway	0–500	1	Unsuitable	
	500–1000	2	Less suitable	
	1000–1500	3	Moderately suitable	
	1500–43 000	4	Highly suitable	
Slope	15–20	1	Unsuitable	
	20–25	2	Less suitable	
	25–30	3	Moderately suitable	
	30–48	4	Highly suitable	
Drainage	0–500	1	Unsuitable	
	500–1000	2	Less suitable	
	000–1500	3	Moderately suitable	
	1500–2000	4	Highly suitable	

10.5.2 Slope mapping analysis

Slope is an important criterion when siting a sanitary landfill site, as it represents the rate of change in elevation at a surface location and is typically expressed in percent or degrees of slope. Areas with a steep slope are not ideal for waste disposal (Kebede & Ayenew, 2023; Zehra *et al.*, 2019). The slope attribute of the study area has been divided into four types. The area suitable for waste disposal is shown in Figure 10.6d in green (class 1).

10.5.3 Drainage mapping analysis

Landfill sites should not be located close to groundwater sources or water assets to prevent contamination of the surface water with leachate. Appropriate buffer distances from meandering and non-meandering rivers must be maintained to prevent leachate infiltrating rivers and major streams. Areas at the lowest elevations, with minimal interaction with groundwater or river levels, is more suitable, as shown in Figure 10.6c: class 1 sites are more suitable for waste disposal (Desta *et al.*, 2025).

10.5.4 Weighted overlay analysis

The weighted overlay analysis (WOA) was completed by overlaying the different thematic raster maps defined above and multiplying each with its given weight and adding them together. The weights were assigned to different thematic grids based on their impact on the environment for site suitability (Arabeyyat *et al.*, 2024; Dande & Mallikarjuna, 2023; Mihiretie, 2022; Ramu *et al.*, 2023).

The weighting of each criterion was determined based on the maximum and minimum values within each layer. All layers were processed and standardized to a common scale of 1, 2, 3, 4: 1 = unsuitable, 2 = less suitable, 3 = moderately suitable, 4 = highly suitable. Finally, all the restrictions were weighted with their respective percentage of influence on an overlay to provide the suitability map. After the overlay analysis of certain features, a suitable location map was created for the disposal of solid waste as shown in Figure 10.7. The output map has four colors (classes), yellow, green, orange, and red. The most suitable area for a waste disposal site is shaded in red (Table 10.2).

10.6 CONCLUSION

The findings of the study demonstrated the capability of GIS and remote sensing technology to systematically identify suitable solid waste disposal sites, thereby minimizing environmental risks and impacts on human health. Such approaches support long-term urban planning and sustainable city development. There is considerable potential for further innovation and the application of advanced technologies in this field.

Selecting landfill sites adhering to the recommendations and regulations of regulatory bodies is a technically challenging issue, but economically beneficial, as it ensures appropriate site selection at the outset, thereby avoiding environmental problems and the need to relocate the site at a later date. The study ensured that all relevant factors and regulations were incorporated through the use of modern technology such as GIS and the weighted overlay analysis method, without wasting time or resources.

The use of GIS, followed by the weighted overlay analysis, highlighted the relative importance of the selected criteria and assisted in assigning appropriate weightage for each criterion. Weighted overlay analysis is one of the most common decision-making techniques used to solve complex, multi-criteria problems.

It is also essential to take into consideration the future increase in the volume of household waste due to rapid population growth and lifestyle changes. Effective strategies for recycling, reuse and long-term planning of landfill sites are necessary, as inadequate planning may lead to many environmental and socio-economic problems at both local and regional levels.

REFERENCES

Abdel-Shafy H. I. and Mansour M. S. M. (2018). Solid waste issue: sources, composition, disposal, recycling, and valorization. *Egyptian Journal of Petroleum*, **27**(4), 1275–1290, https://doi.org/10.1016/J.EJPE.2018.07.003

Aderoju O. M., Dias G. A. and Alberto G. J. (2018). A spaced-based concept in municipal solid waste management and monitoring in developing countries. *International Journal of Environmental Impacts: Management, Mitigation and Recovery*, **1**(4), 482–490, https://doi.org/10.2495/ei-v1-n4-482-490

Ajibade F. O., Olajire O. O., Ajibade T. F., Nwogwu N. A., Lasisi K. H., Alo A. B., Owolabi T. A. and Adewumi J. R. (2019). Combining multicriteria decision analysis with GIS for suitably siting landfills in a Nigerian state. *Environmental and Sustainability Indicators*, **3–4**, 100010, https://doi.org/10.1016/j.indic.2019.100010

Akbari V., Rajabi M. A., Shams R. and Chavoshi S. H. (2008). Landfill site selection by combining GIS and fuzzy multi criteria decision analysis, case study: Bandar Abbas, Iran. *World Applied Sciences Journal*, **3**(1), 39–47.

Aksoy E. and San B. T. (2019). Geographical information systems (GIS) and multi-criteria decision analysis (MCDA) integration for sustainable landfill site selection considering dynamic data source. *Bulletin of Engineering Geology and the Environment*, **78**(2), 779–791, https://doi.org/10.1007/s10064-017-1135-z

Alexakis D. D. and Sarris A. (2014). Integrated GIS and remote sensing analysis for landfill sitting in Western Crete, Greece. *Environmental Earth Sciences*, **72**(2), 467–482, https://doi.org/10.1007/s12665-013-2966-y

Al-Hanbali A., Alsaaideh B. and Kondoh A. (2011). Using GIS-based weighted linear combination analysis and remote sensing techniques to select optimum solid waste disposal sites within Mafraq City, Jordan. *Journal of Geographic Information System*, **03**(04), 267–278, https://doi.org/10.4236/jgis.2011.34023

Al-Hanbali A., Shibuta K., Alsaaideh B. and Tawara Y. (2022). Analysis of the land suitability for paddy fields in Tanzania using a GIS-based analytical hierarchy process. *Geo-Spatial Information Science*, **25**(2), 212–228, https://doi.org/10.1080/10095020.2021.2004079

Alkaradaghi K., Ali S. S., Al-Ansari N., Laue J. and Chabuk A. (2019). Landfill site selection using MCDM methods and GIS in the Sulaimaniyah Governorate, Iraq. *Sustainability*, **11**(17), 4530, https://doi.org/10.3390/su11174530

Aluko O. E. and Igwe O. (2017). An integrated geomatics approach to groundwater potential delineation in the Akoko-Edo Area, Nigeria. *Environmental Earth Sciences*, **76**(6), 240, https://doi.org/10.1007/s12665-017-6557-1

Arabeyyat O. S., Shatnawi N., Shbool M. A. and Shraah A. A. (2024). Landfill site selection for sustainable solid waste management using multiple-criteria decision-making. Case study: Al-Balqa governorate in Jordan. *MethodsX*, **12**, 102591, https://doi.org/10.1016/j.mex.2024.102591

Babalola A. and Busu I. (2011). Selection of landfill sites for solid waste treatment in Damaturu town-using GIS techniques. *Journal of Environmental Protection*, **02**(01), 1–10, https://doi.org/10.4236/jep.2011.21001

Balew A., Alemu M., Leul Y. and Feye T. (2022). Suitable landfill site selection using GIS-based multi-criteria decision analysis and evaluation in Robe town, Ethiopia. *GeoJournal*, **87**(2), 895–920, https://doi.org/10.1007/s10708-020-10284-3

Bilgilioglu S. S., Gezgin C., Orhan O. and Karakus P. (2022). A GIS-based multi-criteria decision-making method for the selection of potential municipal solid waste disposal sites in Mersin, Turkey. *Environmental Science and Pollution Research*, **29**(4), 5313–5329, https://doi.org/10.1007/s11356-021-15859-2

Chang N.-B., Parvathinathan G. and Breeden J. B. (2008). Combining GIS with fuzzy multicriteria decision-making for landfill siting in a fast-growing urban region. *Journal of Environmental Management*, **87**(1), 139–153, https://doi.org/10.1016/j.jenvman.2007.01.011

Collins M. G., Steiner F. R. and Rushman M. J. (2001). Land-use suitability analysis in the United States: historical development and promising technological achievements. *Environmental Management*, **28**(5), 611–621, https://doi.org/10.1007/s002670010247

Dande J. and Mallikarjuna V. (2023). Site selection for solid waste management in Krishna district, Andhra Pradesh using GIS. *E3S Web of Conferences*, **399**, 02001, https://doi.org/10.1051/e3sconf/202339902001

Das G. K. (2020). Site suitability analysis for solid waste disposal of Bolpur Municipality, Birbhum District of West Bengal (India): remote sensing and AHP approach. *Remote Sensing of Land*, **3**(2), 50–58, https://doi.org/10.21523/gcj1.19030201

Demesouka O., Vavatsikos A. and Anagnostopoulos K. (2014). GIS-based multicriteria municipal solid waste landfill suitability analysis: a review of the methodologies performed and criteria implemented. *Waste Management & Research: The Journal for aSustainable Circular Economy*, **32**(4), 270–296, https://doi.org/10.1177/0734242X14526632

Desta M. D., Tesseme T., Yigezu T. T. and Nigussie A. B. (2025). Assessment of landfill site suitability using GIS, remote sensing, and the multi-criteria decision-making (AHP) approach, Ethiopia. *Geology, Ecology and Landscapes*, **9**(2), 662–675, https://doi.org/10.1080/24749508.2023.2256549

Ersoy H. and Bulut F. (2009). Spatial and multi-criteria decision analysis-based methodology for landfill site selection in growing urban regions. *Waste Management & Research: The Journal for a Sustainable Circular Economy*, **27**(5), 489–500, https://doi.org/10.1177/0734242X08098430

Farooqi Z. U. R., Kareem A., Rafi F. and Ali S. (2021). Solid waste, treatment technologies, and environmental sustainability: Solid wastes and their sustainable management practices. In: A. K. Rathoure (ed.). Handbook of Research on Waste Diversion and Minimization Technologies for the Industrial Sector, IGI Global Scientific Publishing, Hershey, PA, USA. pp. 35–57, https://doi.org/10.4018/978-1-7998-4921-6.ch003

Fatoyinbo I. O., Bello A. A., Olajire O. O., Oluwaniyi O. E., Olabode O. F., Aremu A. L. and Omoniyi L. A. (2020). Municipal solid waste landfill site selection: a geotechnical and geoenvironmental-based geospatial approach. *Environmental Earth Sciences*, **79**(10), 231, https://doi.org/10.1007/s12665-020-08973-w

Fraternali P., Morandini L. and Herrera González S. L. (2024). Solid waste detection, monitoring and mapping in remote sensing images: a survey. *Waste Management*, **189**(8), 88–102, https://doi.org/10.1016/j.wasman.2024.08.003

Gautam S., Brema J. and Dhasarathan R. (2020). Spatio-temporal estimates of solid waste disposal in an urban city of India: a remote sensing and GIS approach. *Environmental Technology & Innovation*, **18**, 100650, https://doi.org/10.1016/j.eti.2020.100650

Gbanie S. P., Tengbe P. B., Momoh J. S., Medo J. and Kabba V. T. S. (2013). Modelling landfill location using geographic information systems (GIS) and multi-criteria decision analysis (MCDA): case study Bo, southern Sierra Leone. *Applied Geography*, **36**, 3–12, https://doi.org/10.1016/j.apgeog.2012.06.013

Goel S., Ranjan V. P., Bardhan B. and Hazra T. (2017). Forecasting solid waste generation rates. In: D. Sengupta and S. Agrahari (eds). Modelling Trends in Solid and Hazardous Waste Management, Springer, Singapore, pp. 35–64, https://doi.org/10.1007/978-981-10-2410-8_3

Gorsevski P. V., Donevska K. R., Mitrovski C. D. and Frizado J. P. (2012). Integrating multi-criteria evaluation techniques with geographic information systems for landfill site selection: a case study using ordered weighted average. *Waste Management*, **32**(2), 287–296, https://doi.org/10.1016/j.wasman.2011.09.023

Hailu Y, Hanchiso T. and Bereta A. (2019). Municipal solid waste suitable disposal site selection, case study, Wolkite Town, Ethiopia. *International Journal of Environmental Sciences & Natural Resources*, **20**(4), pp.126–131. https://doi.org/10.19080/IJESNR.2019.20.556044

Haripavan N. and Dey S. (2023). Application of remote sensing and geographic information system in solid waste management for Gudivada Municipality, Andhra Pradesh, India. *Waste Management Bulletin*, **1**(3), pp.128–140. https://doi.org/10.1016/j.wmb.2023.08.006

Irhoumah J. M. S., Agarwal V. C., Lal D. and Kumar M. (2014). Determination of suitable site for solid waste disposal using remote sensing and GIS techniques in Allahabad Municipality Area. *International Journal of Engineering Research and Technology*, **3**(6), pp.876–892. https://doi.org/10.17577/IJERTV3IS060921

Kebede H. A. and Ayenew W. A. (2023). Solid waste disposal site selection analysis using geospatial technology in Dessie city Ethiopia. *Heliyon*, **9**(8), e18865, https://doi.org/10.1016/j.heliyon.2023.e18865

Khan D. and Samadder S. R. (2014). Municipal solid waste management using geographical information system aided methods: a mini review. *Waste Management and Research*, **32**(11), 1049–1062, https://doi.org/10.1177/0734242X14554644

Khan D. and Samadder S. R. (2015). A simplified multi-criteria evaluation model for landfill site ranging and selection based on AHP and GIS. *Journal of Environmental Engineering and Landscape Management*, **23**(4), 267–278, https://doi.org/10.3846/16486897.2015.1056741

Kontos T. D., Komilis D. P. and Halvadakis C. P. (2005). Siting MSW landfills with a spatial multiple criteria analysis methodology. *Waste Management*, **25**(8), 818–832, https://doi.org/10.1016/j.wasman.2005.04.002

Mallick J. (2021). Municipal solid waste landfill site selection based on fuzzy-ahp and geoinformation techniques in ASIR region Saudi Arabia. *Sustainability (Switzerland)*, **13**(3), 1538, https://doi.org/10.3390/su13031538

Mihiretie A. (2022). Suitable solid waste disposal site selection using geographical information system: a case of Debre Markos Town, Ethiopia. *International Journal of Environment and Geoinformatics*, **9**(2), 1–8, https://doi.org/10.30897/ijegeo.784087

Mussa A. and Suryabhagavan K. V. (2019). Solid waste dumping site selection using GIS-based multi-criteria spatial modeling: a case study in Logia town, Afar region,

Ethiopia. *Geology Ecology and Landscapes*, **5**(3), 186–198, https://doi.org/10.1080/24749508.2019.1703311

Okot P., Ogao P. J. and Abandu J. (2019). Site selection model for urban solid waste disposal management using GIS and remote sensing: a case of Gulu Municipality. *International Journal of Environment and Waste Management*, **24**(4), 405–436, https://doi.org/10.1504/IJEWM.2019.103645

Perdue W. C., Stone L. A. and Gostin L. O. (2003). The built environment and its relationship to the public's health: the legal framework. *American Journal of Public Health*, **93**(9), 1390–1394, https://doi.org/10.2105/AJPH.93.9.1390

Ramu P., Santosh B. S. and Praveen S. (2023). An integrated GIS-AHP approach for municipal solid waste landfill siting in Srikakulam District, Andhra Pradesh. *Nature Environment and Pollution Technology*, **22**(1), 139–148, https://doi.org/10.46488/NEPT.2023.V22I01.012

Rezaeisabzevar Y., Bazargan A. and Zohourian B. (2020). Landfill site selection using multi criteria decision making: influential factors for comparing locations. *Journal of Environmental Sciences*, **93**, 170–184, https://doi.org/10.1016/j.jes.2020.02.030

Saketa Y., Tamene N. and Melknew M. (2023). Municipal solid waste disposal site suitability analysis using multi-criteria evaluation in Assosa, Ethiopia. *International Journal of Environmental Science and Technology*, **20**(4), 3815–3830, https://doi.org/10.1007/s13762-022-04221-9

Sanjay S., Nalina M. E. and Ajith Hebbar H. (2018). RS and GIS based site suitability analysis for the disposal of solid waste for Moodbidri and Karkala region. *International Journal of Civil Engineering and Technology*, **9**(5), 594–601.

Santhosh L. G. and Sivakumar Babu G. L. (2018). Landfill site selection based on reliability concepts using the DRASTIC method and AHP integrated with GIS–a case study of Bengaluru city, India. *Georisk: Assessment and Management of Risk for Engineered Systems and Geohazards*, **12**(3), 234–252, https://doi.org/10.1080/17499518.2018.1434548

Şener B., Süzen M. L. and Doyuran V. (2006). Landfill site selection by using geographic information systems. *Environmental Geology*, **49**(3), 376–388, https://doi.org/10.1007/s00254-005-0075-2

Şener Ş., Sener E. and Karagüzel R. (2011). Solid waste disposal site selection with GIS and AHP methodology: a case study in Senirkent–Uluborlu (Isparta) Basin, Turkey. *Environmental Monitoring and Assessment*, **173**(1–4), 533–554, https://doi.org/10.1007/s10661-010-1403-x

Sumathi V. R., Natesan U. and Sarkar C. (2008). GIS-based approach for optimized siting of municipal solid waste landfill. *Waste Management*, **28**(11), 2146–2160, https://doi.org/10.1016/j.wasman.2007.09.032

Tadesse T. (2004). Lecture notes: solid and hazardous waste management.

Vergara S. E. and Tchobanoglous G. (2012). Municipal solid waste and the environment: a global perspective. *Annual Review of Environment and Resources*, **37**, 277–309, https://doi.org/10.1146/ANNUREV-ENVIRON-050511-122532

Yal G. P. and Akgün H. (2013). Landfill site selection and landfill liner design for Ankara, Turkey. *Environmental Earth Sciences*, **70**(6), 2729–2752, https://doi.org/10.1007/s12665-013-2334-y

Zehra S., Afsar S. and Anis Y. (2019). Municipal solid waste disposal site selection in Jacobabad using RS/GIS techniques. *Journal of Geoscience and Environment Protection*, **07**(10), 13–22, https://doi.org/10.4236/gep.2019.710002

Zeleke J. A. and Wedajo G. K. (2025). Geospatial technologies and multi-criteria decision making based-urban solid waste disposal suitable site selection: the case of Assosa town, Ethiopia. *Cleaner Waste Systems*, **11**, 100284, https://doi.org/10.1016/j.clwas.2025.100284

Index

IWA Publishing's authorised EU representative for General Product Safety Regulations is Diane D'Arras, 15 rue Duret, 75116 Paris, France, e-mail: safety@iwap.co.uk.

Printed and bound by CPI Group (UK) Ltd, Croydon, CR0 4YY
13/10/2026
02232602-0001